Copper Mining
in
Santa Rita, New Mexico
1801–1838

Copper Mining in
Santa Rita, New Mexico
1801–1838

A New Mexico Centennial History Series Book

Helen Lundwall
with
Terrence Humble

SUNSTONE PRESS

SANTA FE

Sunstone books may be purchased for educational, business, or sales promotional use.
For information please write: Special Markets Department, Sunstone Press,
P.O. Box 2321, Santa Fe, New Mexico 87504-2321.

Book and Cover design › Vicki Ahl
Body typeface › Franklin Gothic Book
Printed on acid-free paper
∞

Library of Congress Cataloging-in-Publication Data

Lundwall, Helen J., 1926-
  Copper mining in Santa Rita, New Mexico, 1801-1838 / by Helen Lundwall
with Terrence Humble.
     p. cm. – (A New Mexico centennial history series book)
  Includes bibliographical references and index.
  ISBN 978-0-86534-888-2 (softcover : alk. paper)
 1. Copper mines and mining–New Mexico–Santa Rita–History–19th century. 2. Mining camps–
New Mexico–Santa Rita–History–19th century. 3. Copper miners–New Mexico–Santa Rita–
History–19th century. 4. Santa Rita (N.M.)–History–19th century. 5. Santa Rita (N.M.)–
Economic conditions–19th century. 6. Frontier and pioneer life–New Mexico–Santa Rita.
7. Santa Rita (N.M.)–Social life and customs–19th century. I. Humble, Terrence M., 1941-
II. Title.
  TN443.N6L86 2012
  622'.34309789692–dc23
                              2012015913

WWW.SUNSTONEPRESS.COM
SUNSTONE PRESS / POST OFFICE BOX 2321 / SANTA FE, NM 87504-2321 /USA
(505) 988-4418 / ORDERS ONLY (800) 243-5644 / FAX (505) 988-1025

In Memory of

*A*ficionados of Southwest History
Max LeRoy Lundwall and Patrick M. Humble

Contents

Foreword

This is the story of a fabulous ore body. It was first commercially developed in 1801 and is still being profitably exploited today over 200 years later. This mine has a fascinating history and a longer working life than any copper mine in United States history.

In her work, Helen Lundwall writes about the first development of the Santa Rita copper mines by the early Spanish colonizers. This book is thoroughly researched, well written and is the definitive record of the first chapter in the long history of the exploitation of this richly mineralized area in southwestern New Mexico.

But, it is a story of only the beginning. The Santa Rita del Cobre mine, as it was commonly known, grew over the years into a great mining industry which supplied a significant amount of the copper necessary for our country to become the economic leader of the world.

After the miners were forced to abandon the area in 1838, Santa Rita remained a virtual ghost town for a number of years. By 1872 the Apaches were subdued and various American entrepreneurs attempted to profitably mine the remaining high-grade veins. These high-grade veins began to play out into massive sulphide rocks that contained too little copper to warrant mining. At the time, no technology was available to recover this low-grade copper.

In the early 1900s, technology was developed, first in Utah and then at Santa Rita, to recover this low-grade copper. In 1909, the Chino Mines Company was formed. A young engineer named John M. Sully was the first general manager. In 1910 the first steam shovel began work at the mine and in 1911 a mill was built in Hurley to concentrate the ore. A railroad then transported the ore nine miles to the Hurley mill.

In the following years, Chino Mines grew at a rapid rate. Kennecott Copper Corporation acquired the property in 1933 and built a smelter, also in Hurley. The capital expenditure of hundreds of millions of dollars for ever-improving technology over succeeding years allowed the company to continue to operate. First, steam was replaced by electricity as a means of powering the giant shovels, drills, and the full-scale railroad trains used in mining. Then the trains were replaced with large off-highway trucks. The trucks, shovels and drills became bigger and more efficient. Today (2012), 290 ton trucks and 56-cubic yard shovels are operating at the mine.

During the period from 1981 to 1985, $450 million in additional capital was employed to again improve the operation. The mining capacity was increased at the mine, as a new state-of-the-art computer-controlled mill was built which replaced the old smelting technology with an oxygen flash furnace.

The ownership changed again in 1987 when Phelps-Dodge Corporation assumed control of the property. This organization continued to install improved technology to process the ever-decreasing ore grade. A solvent extraction/ electro-winning facility was built in order to recover the very low-grade copper in the ore types that could not be treated by concentrating and smelting.

In March of 2007 the ownership changed once more. Freeport-McMoRan, a mining company from Louisiana, bought the operations of Phelps Dodge Corporation, including Chino Mines Company. It would appear that the mine is under enlightened and forward-looking management, keeping in line with the management styles of previous owners.

Over this long period of time, Chino Mines has provided the basis for a fine living for thousands of employees and their families. It has been one of the largest private employers in New Mexico and has been an economic mainstay in the southwest part of the state. It will remain a successful mining operation in the future if technology improvement and employee productivity increases continue to offset declines in the copper content of the ores.

This book tells us how this fabulous mining saga got its start. It also vividly describes an earlier epoch in the history of mining in the American West. I now invite the reader to enjoy this history that has fascinated us all.

—David A. Kinneberg
Former General Manager
Chino Mines Company

Preface

*C*opper Mining in Santa Rita, New Mexico, 1801–1838 is the story of the formative years of a remarkable mine in southwestern New Mexico that has produced copper for more than 200 years. Records of the Spanish Colonial and early Mexican period yield intriguing accounts of the people involved in the early development of the mines, the difficulties they encountered along the way, and the importance of this small settlement to the history of the frontier.

The middle of Apache homelands, 150 miles from the nearest Spanish settlement, seems a most unlikely place to find a successful mining operation. No one knew how the aggressive Apaches, who inhabited much of southern New Mexico and Arizona, would react to a Spanish settlement in their midst. These Apaches caused many problems and had a strong influence over the success or failure of the Santa Rita mining operation. At times the hostility and depredations of these Indians overshadowed the remarkable success of the mines. Over the years, the mining camp became the center of operations in the war against the Apaches. The governor of Chihuahua called Santa Rita del Cobre, as it was known, "the watchtower and guardian of the western frontier."[1]

Santa Rita copper played an important role in the economic development of New Spain's northern frontier. The Sierra del Cobre mines, or El Cobre as Santa Rita was first known, had a seemingly inexhaustible store of exceptionally fine copper. An assay of Santa Rita copper at the Mexico City mint in 1804 established the fine quality of this metal. It was very pure, malleable and easy to refine, "the best copper that has come into this mint in many years." The mint was ordered to pay top dollar (25 pesos per hundred weight) for all the copper these mines could produce. The Santa Rita mines prospered through the last years of Spanish Colonial rule. The mine survived a decade of revolution with few scars. As Mexico struggled to establish a stable government in the early years of independence, Santa Rita continued to turn out high-grade copper. During these often difficult years, there was always a need for copper. When one market faded, another appeared. Although the Santa Rita mines yielded a fortune to the few men willing or able to invest money in their development, it was always a difficult and hazardous undertaking. History professor Paige Christiansen stated, "by far the most interesting copper mine in the United States is the old Santa Rita mine . . ."[2]

This book is more than the history of a mine. It is the story of people of courage and determination who developed a thriving business and established homes in this out-of-the-way place. It portrays life as it was for the mineworkers and their families. These were tough, hardy pioneers who lived a harsh and unforgiving life. The miners worked long, hard hours under dangerous conditions. Their families endured hardship and privation. Mule trains brought in all the necessities of life–food, clothing, and mining supplies–from distant places. The mined copper was transported to market in the same manner. There was always the fear of a deadly raid from hostile Apaches. Many times, the future of Santa Rita was in doubt in these early years as the residents adapted, improvised and endured. This is their story.

Acknowledgments

We are proud to have a place in the Public History Program at New Mexico State University to commemorate the history and heritage of New Mexico on the anniversary of statehood. We could not ask for a better circumstance to present the 200 year-old story of the first settlement in the far southwestern part of our great state. Our thanks go to Jon Hunner, Director of the Public History Program at New Mexico State University in Las Cruces, New Mexico.

I am also grateful to Terrence Humble. His proficiency and diligence in translating documents from the Spanish and Mexican archives made it possible to retrieve much of the story of mining in Santa Rita. He also provided the epilogue. We are deeply grateful to our families for their support and encouragement: Artemisa Humble for help with the Spanish language; Lee Lundwall for his assistance in locating and interpreting early maps of the Santa Rita area; Linda Brake for proofreading and editorial assistance.

We are indebted to the following people whose material assistance, advice and encouragement were of great help: Dave and Liz Kinneberg, Thomas Brake, Ellen Cline, Phillip Parotti, Dr. Myra Ellen Jenkins, formerly of the New Mexico Record Center and Archives; Annie H. Jordan, librarian of the Nettie Benson Latin American Collection, University of Texas at Austin; Cesar Caballero, Director of Special Collections, University of Texas at El Paso; the late Francisco R. Almada, historian of Chihuahua, Mexico, who generously shared information from his own files; Robert Eveleth, New Mexico Bureau of Mines, Socorro, New Mexico; William Griffen, Northern Arizona University; Anthony Romero, local historian of the Janos Trail; Susan Berry and Jacquiline K. Becker of the Silver City Museum; Steve Hussman and Dean Wilkey, Archives and Special Collections, New Mexico State University, Las Cruces; George Austin (Silver Imaging). Without their assistance, this project would not have come to fruition.

1

Apachería

The Apaches weave like a crimson thread through the story of Santa Rita del Cobre, as the mine was known. Of all the Indians across New Spain's northern frontier, and there were many, the Apaches of southwestern New Mexico and southern Arizona were the most destructive and indomitable. They made a significant impact on the economic development of New Spain's northern frontier.

Santa Rita del Cobre is located at the south end of the Mimbres mountains of southwestern New Mexico, near present-day Silver City. In the eighteenth century Spanish explorers named this richly mineralized area *Sierra del Cobre Virgen* (mountain of native copper). This was the heart of Apachería, a far-reaching territory extending from the central Rio Grande across southwestern New Mexico and southern Arizona. From high, timbered mountains, deep canyons, and grassy meadows, to fertile river valleys and barren desert land, Apachería was home to numerous bands of Apaches who found the environment well suited to their way of life. They were hunters and gatherers, with a propensity for raiding New Spain's frontier settlements.

The nomadic Apaches migrated from western Canada, reaching the southwest about the same time the Spanish conquestadors came to colonize New Mexico. Driven from buffalo country by the stronger Plains Indians, the Apaches scattered over western Texas, New Mexico and eastern Arizona, establishing their *rancherías* in natural fortifications in the rugged mountains, where water and wood were available and game was plentiful. A ranchería was a cluster of brush huts occupied by family members and extended family. Several rancherías with common interests might form a band for security and efficiency in a cooperative activity such as raiding, warfare, hunting and food gathering.[1]

As Spanish civilization moved into northern Mexico and up the valley of the Rio Grande, raiding the New Mexico settlements and pueblos for food and animals became an important economic activity for the Apaches. The Indians acquired horses from the Spanish conquerors, and soon became proficient horsemen. They widened their field of operation to include northern Chihuahua and Sonora. Several bands, operating simultaneously, could strike swiftly and unexpectedly, burning villages and isolated haciendas, destroying crops, stealing livestock, foodstuffs and other plunder. They captured women and children to adopt into their tribe, and killed many an unwary traveler who ventured along the byways. If pursued by the military, the Indians disappeared into the formidable mountains and deep canyons of Apachería, not to be seen again until the time was right for the next raid. These incessant depredations played havoc with the development of New Spain's northern borderlands.[2]

Spanish authorities tried to halt the terror and destruction laying waste to the frontier, and bring these aggressive Indians under control. They sent large military forces into Apachería to explore the country and seek out and destroy the enemy. Although punitive operations against the wily Apaches were seldom successful, the Spaniards did become more familiar with Apachería and the Indians who inhabited this sizeable piece of land on the fringes of New Spain. They applied names to the various Apache bands according to the particular areas they inhabited. The groups more closely tied to the history of Santa Rita del Cobre were the *Mimbreños*, whose homeland was the Mimbres Mountains and along the Mimbres River, and the *Mogollónes,* whose territory centered around the Mogollon Mountains and the upper Gila River. Their primary target was Chihuahua and the Rio Grande settlements, while to the west were *Gileños* and *Chiricahuas*, who usually raided Sonoran settlements.[3]

Bernardo de Miera y Pacheco, a militia captain from El Paso del Norte, rode on one of the earliest Spanish expeditions through Apachería in 1747. As an engineer and skilled cartographer, Miera had the task of mapping the *tierra incognita* (unknown land) of Apachería. No doubt he took note of prominent landmarks along the way, including the Sierra del Cobre, the future site of the Santa Rita mines. Information gathered by Miera on this and other expeditions appeared in his 1758 map of New Mexico commissioned by Governor Marin del Valle.[4]

The interest of cartographer Bernardo de Miera in the Sierra del Cobre area surfaced again in 1777, when Apache warfare was at its peak. He sent King Charles III of Spain a proposal for dealing with the hostile Apaches which called for establishing settlements in their midst and reaping the profits from the mines in their mountains. Miera envisioned Apachería as the province of *Jila* (Gila). From three outposts, a strong military force would wage continuous war against the enemy Apache and dislodge them from the new province. According to Miera's plan, Paso de Todos los Santos, on the Gila River near present-day Cliff, New Mexico, presented an ideal site for the first colony. He recommended the valley of the Mimbres River near the Sierra del Cobre Virgen for the second settlement, with the third located near San Pasqual on the Rio Grande. Miera, aware that the Sierra del Cobre contained a treasure of valuable minerals, volunteered to command the second outpost on the Mimbres. He promised to have the three settlements established, mines of gold, silver, and copper located and in operation, and peace restored to the area within three years.[5] Miera's remarkable plan did not gain the approval of King Charles.

Spain's North American colony, claimed by right of discovery, extended from the Mississippi to the Pacific coast. She needed strong borders and established settlements to act as buffers against encroachment from rival European countries seeking a foothold in North America. Spain had already lost possessions in the Caribbean and South America, and her hold on Florida and Louisiana appeared tenuous. California was sparsely settled and poorly defended, vulnerable to a takeover from the sea. It was imperative to open land routes to the west coast for trade and strategic purposes.[6] The most serious impediment to progress came from the Indians of Apachería, whose frequent depredations brought northward expansion to a halt.

An inspection of the borderland in 1777 revealed the sorry state of the frontier provinces. The Apaches had done irreparable damage to the lands bordering Apachería. Between 1771 and 1776, it was reported that in the province of Nueva Vizcaya (Chihuahua and Durango), 1,674 persons were killed and 154 captured in Apache raids; 116 haciendas and ranches were abandoned and 68,256 horses and mules stolen. The value of property destroyed or stolen exceeded twelve million pesos. In 1771 alone the Apaches stole seven thousand horses and mules and destroyed many herds of cattle and flocks of sheep. By the middle of the eighteenth century, Apache depredations reached critical proportions as settlers deserted villages and ranches, and miners abandoned mines that held great promise.[7]

From the Spanish point of view, the long-standing conflict with the Indians was a war. They considered the Apache a ruthless and savage predator, a menace to the lives and properties of frontier pioneers. The Indians, however, wanted only to preserve their way of life. In their eyes, plundering and terrorizing the frontier settlements was not an act of war aimed at increasing their territory or defeating the Spanish in battle. They raided from economic necessity in order to obtain food and horses. Sometimes they sought revenge for wrongs which they thought the Spaniards committed against them. Livestock and booty taken in raids enhanced a warrior's reputation as a fighter. With bitterness and hatred on both sides, it was difficult to find a viable solution.[8]

It was futile to try to subdue the raiders through military action alone. The Spanish military operations were mostly ineffective. The *presidios* (forts) were undermanned and a great distance apart, allowing the Apaches to slip through and raid the interior villages. The supply system often broke down, leaving the soldiers without food, clothing, and payroll. The troops were undisciplined and untrained in dealing with the hit-and-run tactics of the Apaches. There were times when the brazen Indians ran off entire presidial horse herds, leaving the soldiers afoot for many weeks.[9]

The Spanish instituted a new Indian policy that consisted of aggression followed by peace agreements. The position of *Comandante General de Provincias Internas* (Commandant General of the Northern Provinces) was created to carry out the new policy. Under his direction, the entire frontier military system was reorganized. New presidios were established and old posts were moved to more strategic locations, manned by additional well-trained and diciplined soldiers.[10]

Beginning in the late 1770's, well-planned military campaigns, coordinating the combined forces of several presidios, civil militias, Indian auxiliaries and Indian scouts, carried the battle deep into Apache territory. They invaded the Apache's strongholds, destroying rancherías, killing or taking warriors and their families prisoner, and recovering stolen animals and property. At first there was little improvement in the situation. The Apaches continued their depredations, now seeking revenge for the loss of relatives and possessions. In response, the Spanish military intensified their patrols and offensive operations.[11]

These coordinated offensives eventually produced the desired results. Many Apache groups came to the presidios to negotiate for peace. The terms of the peace treaties gave Spanish authorities strict control over the activities of the Indians. The Apaches agreed to give up their barbaric ways and cease harassing frontier settlements. They moved their rancherias to reservations (referred to as peace establishments) near a presidio where the military commander had jurisdiction over them. At certain times of the year *guías* (travel permits) were issued which allowed them to return to their homeland to hunt and gather wild plants in their season. The families received weekly rations of corn or wheat, meat, brown sugar, salt and tobacco; they were given tools and a plot of ground on which to raise their own food in order to become self-sufficient, although this was never entirely successful. The treaty allowed the Apaches to keep the horses and livestock already in their possession, but forced them to return all captives.[12]

Through this policy of military aggression followed by peace agreements, the Spaniards brought some measure of peace to the northern frontiers. By 1793 some two thousand Apaches lived at peace establishments near designated presidios. The population was not always static, however. The settled warriors missed the excitement of raiding and eluding their pursuers. They were often restless and required close supervision by the military. At times rations were scarce and some bands left the peace establishments in order to return to their home territory.[13]

Not all Apaches accepted the restrictions of a truce. An estimated two-thirds of the Apache bands negotiated for peace at one time or another, while the remainder refused to give up their freedom. Several hostile bands still inhabited the southern mountains of New Mexico and Arizona and continued to raid the frontier settlements.[14] Nevertheless, Spanish-Indian relations improved significantly enough for the Spanish government to consider it a time of peace, and the economy began to show signs of recovery. Owners reclaimed abandoned haciendas, and agriculture and stock raising began to flourish. With the encouragement of the government, miners and prospectors renewed their interest in mining, reactivated old workings and located new ore deposits.

In this interim of peace Josè Manuel Carrasco began to develop the first copper mine at Santa Rita del Cobre. For many years to come, however, the Apaches would have an active role in the success or failure of these copper mines.

2

Carrasco Stakes His Claim

José Manuel Carrasco, a retired Lieutenant Colonel of the Spanish Colonial army, was responsible for developing the first copper mine in the area now known as Santa Rita. It required an extra measure of courage and determination to exploit the rich mineral resources found in the heart of Apachería. Carrasco encountered so many problems when he tried to capitalize on his discovery that he wondered if he could ever make a profit from his new mines.

Carrasco was born in 1743 at San Antonio de Julimes, Chihuahua, Mexico, and grew up on the frontier. He joined the colonial army of New Spain at the age of twenty-five, and served at the frontier presidios of Carrizal, Janos, and San Buenaventura, participating in many expeditions against the Apaches during his long military career. Several times he received citations for bravery in action when leading his company against hostile Indians. For example, in October 1777 Apaches tried to steal the entire horse herd from the presidio of Janos. Quick action by Carrasco drove the raiders off and saved the animals.[1] At another time, he met an Apache band four times the size of his own command; he defeated the enemy and won a promotion to the rank of captain. Three years later he became commander of the *Primera Compañía Volante*, a mobile troop of mounted soldiers ready to pursue raiding Indians anywhere in the province. Lieutenant Colonel Carrasco retired from the military service in 1798, after thirty years of active duty, and established his home at Janos.[2]

While stationed at the presidios of Janos and Carrizal, Carrasco made friends with Apaches living in nearby peace establishments. He often spent his own money to provide clothes and other necessities for indigent Apaches, thus gaining the good will of these Indians.[3] In 1799 Carrasco's Apache friends offered to take him to a mountain in Apache lands far to the north where they said there was gold. A multitude of Apaches still inhabited Apachería.

Some bands who made peace with the Spaniards were allowed to live in their homeland as long as they maintained that peace. Other Apacheríans never surrendered, and continued to harass the settlements of the northern frontier. Carrasco fought against many of these Indians during his military career. He might have felt some qualms about the danger he would face in this venture, but the old soldier's pension was small and the lure of gold was strong.

After a journey of several days, Carrasco and his Apache guides reached the foothills of the Sierra del Cobre.[4] They followed a fair-sized creek up a canyon into a small valley which widened into a basin of several hundred acres surrounded by a high ridge of hills. On the east, the ridge rose to a height of sixteen hundred feet, topped by a peculiar crest of bare rock which ended in a sharp bluff. A column of rock, vaguely resembling a human figure, stood tall before the bluff. The Spaniards called this prominent monolith *la bufa*; today it is a famous landmark known as the Kneeling Nun.[5] Through the years, the process of weathering and earth tremors have reduced the nun to a smaller figure, but it remains a treasured memento of the past.

The Kneeling Nun. This curious rock formation on a bluff overlooking the Santa Rita basin was a well-known landmark to soldiers, miners, prospectors, trappers and traders traveling in Apacheria. Humble collection.

An amazing sight greeted Carrasco as he approached the area. Scattered across the floor of the basin were large outcrops of white porphyry liberally laced with copper. The porphyry matrix, exposed by weathering, was soft and easily separated from the pure metal which appeared in fractures and cavities in the rock and ranged from thin flakes and pellets to slabs half an inch thick and several feet wide.[6] Carrasco came to this precarious place seeking gold, but it was the abundant evidence of easily obtained copper that led him to stake two claims on the rich red metal that he thought would bring a sizable profit. He also located claims of gold and silver at this time, but never had the resources to develop all his discoveries.

The time was right for copper miners. Spain, involved in endless European wars throughout the eighteenth century, had an urgent need for Mexican copper. Her armament factories required large amounts of copper to make cannon, small arms, ship fittings, and armature. To obtain an adequate supply of this metal, Spain turned to her colonies in the new world. Spanish authorities ordered Mexico to ship a specified amount of copper to Spain each year; Mexico's fledgling copper industry could not meet these quotas. In 1780 Spain established a government monopoly that required Mexican miners to sell all the copper they produced to the royal treasury in Mexico City. The treasury paid eighteen pesos per *quintal* (one hundred pounds) of copper. The expense of removing the ore, getting it to market, and paying the required fees and taxes left the mine owner with little profit. Prospectors and miners had no incentive to discover and open new mines or expand those already in operation. To put new life into the business of producing copper, the viceroy raised the price to twenty-two pesos per hundred pounds in 1799. This concession doubled the production of Mexican copper within two years, and probably influenced Carrasco in his decision to stake his claim on the copper of the Sierra del Cobre.[7]

The Sierra del Cobre had excellent prospects for becoming an important new mining district. A fair creek, fed by numerous springs, flowed through the area. Juniper, scrub oak and tall pines covered the hills around the mineral discovery, providing a ready supply of building material and fuel. Deer, antelope, quail, turkey and bear inhabited the surrounding country. Grama and other grasses grew tall, providing excellent pasture for horses, cattle and sheep. Although the basin itself provided only a few acres for cultivation, there was abundant farmland in the nearby valley of the Mimbres River.[8] It was more than a year, however, before Carrasco could do the required assessment work, register his claims with the *Deputatión de Minería* (mining tribunal) and take possession of his Sierra del Cobre discovery.

The logistics encountered in opening mines in this out-of-the-way place presented difficulties. Carrasco had to resolve four major problems: operating capital, labor, transportation, and hostile Apaches. He had to recruit miners, transport them to the site, protect them from the Indians, and furnish all the food, clothing, supplies, and tools needed at the mine.[9] He also faced the difficult and costly task of getting the ore to the royal treasury in Mexico City, and paying the taxes and fees required there.

Carrasco had no money. He needed a backer willing to make a sizable investment in this risky venture. Prosperous merchants often invested in promising mines by providing money and supplies, securing their investment with a mortgage on the ore produced or a partnership in the discovery. Sometimes the merchant acted as ore buyer, purchasing the metal at a discounted rate and transporting it to the royal treasury.[10] Chihuahua merchant Pedro Ramos de Verea was willing to back Carrasco in the initial development of his Sierra del Cobre mines and act as his agent in Chihuahua. Pedro Ramos was one of several merchants supplying the frontier presidios; he might have solved Carrasco's transportation problem by using his own mule-trains to haul the ore to Chihuahua where he made arrangements for further shipment to Mexico City.

Labor presented a problem because of the difficulty of finding men willing to work in a place as isolated and dangerous as the Sierra del Cobre. Experienced miners wanted twice their regular wage to undertake such a hazardous venture. Carrasco appealed to Don Pedro de Nava, Commandant General of the Internal Provinces, for assistance.[11] Nava rounded up a labor force of peons, petty criminals, and Apaches from the peace establishments. It was common practice to assign petty criminals to mines and haciendas to work out their debts. According to Carrasco, he paid the Mexicans one peso per day in wages, and the Indians received one-half peso.[12] It is not clear what role the Apaches played; were they along to labor in the mine, or to assure the safety of the party?

In the fall of 1800, as Carrasco prepared to take his crew into the Sierra del Cobre, he discovered one more problem. Hostile Apaches still occupied areas of the Mimbres valley and around San Vicente Springs (present-day Silver City). Given the uncertain temperament of the renegades, the threat was too near the site of Carrasco's proposed workings. Commandant General Nava ordered soldiers from Janos and San Buenaventura to make a reconnaissance of the Mimbres Mountains and the Sierra del Cobre and drive out any Indians found in the vicinity.[13] In the spring of the following year, the troops made another sweep of the area, forcing the Indians to move into the distant mountains.

Carrasco, with his motley crew and the approval of the Commandant General, finally returned to Apachería in April 1801 to begin work on his mining claims. Two months later his agent, Pedro Ramos, registered Carrasco's Sierra del Cobre discoveries with the Deputatión de Minería in Chihuahua: two veins of copper (*El Corazon de Jesus* and *El Corazon de María*), a silver claim (Guadalupe), and a gold placer (*Santisima Trinidad*). He presented a small sample of gold from the placer and a specimen of native copper weighing four pounds eleven ounces to verify the presence of these minerals.[14] The Crown owned all mineral wealth in its domain, but individuals could discover and work mineral deposits if they registered the mines and paid the Crown a royalty on their production.[15] The mining laws stipulated that assessment work would be done within ninety days in order to take possession of a new claim. Because of the great distance from Chihuahua and the danger from Indians in the area, this time limit was waived through 1806 for claims registered in the Sierra del Cobre.[16]

Carrasco's new mines were in the province of New Mexico, some three hundred miles from the seat of government at Santa Fe. Troops from the presidio of Janos, one hundred fifty miles to the south, often visited the Sierra del Cobre when on routine patrol of Apachería. The Commandant General decided to place the new mining area under the jurisdiction of Janos where it would receive more support and protection. Records and correspondence of this period first referred to the new discovery as the Sierra del Cobre mines or Minería del Cobre. As their fame increased, the mines became known simply as El Cobre.

Carrasco needed only a rudimentary knowledge of mining to develop his copper claim. The copper, exposed by erosion, appeared in a multitude of cracks and crevasses in soft, white porphyry on or near the surface.[17] Unskilled laborers worked the surface ores from open trenches and shallow pits. It was a simple matter to break up the soft, friable base stock with a heavy iron crowbar and extract the malleable native copper. These primitive mining methods produced all the rich ore that Carrasco could handle.

By July 1801 Carrasco was ready to take his first shipment of 84 *arrobas* (2,127 pounds) of ore to Chihuahua. The rough pieces of pure copper came in all sizes and shapes. He packed the ore in rawhide bags, loaded the cargo on eight mules, and joined the Janos supply train for the journey south. In Chihuahua his agent made arrangements to transport the ore to the royal treasury in Mexico City. It is unlikely that Carrasco saw much profit from this first shipment of copper. Captain José Manuel Ochoa, commander of Janos presidio, claimed two loads of copper.[18] Perhaps Ochoa invested money in Carrasco's mining venture or claimed payment for escort service of the Janos troops. Any money remaining after Carrasco settled his debt with Pedro Ramos went toward financing the next trip to the mines.

Carrasco did not attempt to establish a permanent camp at his Sierra del Cobre mines. The new mines were situated in a precarious position in the midst of Apache lands. His small workforce could not defend itself against the Indians who made their presence known every now and then. The workers refused to remain at the mine when Carrasco took the copper to Chihuahua. He had no choice but to take the crew back and forth with him on each trip. He asked the Commandant General for an escort of Ópata Indians. These sturdy, well-disciplined Indian troops from the company at Bavispe, Sonora, were invaluable to the army. They were good soldiers, excellent marksmen, and were able to travel long distances by living off the land.[19] Unfortunately, the Ópatas had such a menacing and warlike appearance that Carrasco's workers were afraid of them. After making eight trips with him, the labor force refused to return to the Sierra del Cobre without a regular military escort.[20]

In the next two years, Carrasco made several shipments of copper from his mine, but he realized very little profit. As he explained to the Deputación de Minería, the great distance from the mines to Chihuahua City resulted in high transportation costs. He paid for smelting and refining the ore plus the *diezmo* (Royal tithes). There were additional expenses for freight charges to transport the refined copper to Mexico City and the six percent tax charged there. He found it impossible to make enough money to continue his mining venture. Again Carrasco took his problems to the Commandant General, who agreed to temporarily waive the tax levied by the government.[21]

Pedro de Nava turned the office of Commandant General over to Brigadier General Nemecio Salcedo[22] in November of 1802 and returned to Spain. With the northern frontier provinces now experiencing more peaceful relations with the Indians, the new Commandant General could focus his attention on restoring the frontier to economic prosperity. The mining industry received his full support. He encouraged prospecting for new ore deposits, furnished military escorts when needed and granted concessions to miners in order to promote the business. In spite of the dangerous location of the Sierra del Cobre mines, Salcedo decided they might prove worthy of his attention. With the strong backing from the Commandant General, others dared to join Carrasco at the place now called El Cobre.

Francisco Javier Bernál and Francisco García de Noriega, residents of El Paso del Norte, heard of the discovery of copper in the Sierra del Cobre. In midsummer of 1803, they came to look over Carrasco's mine and investigate the prospects of the area. They saw evidence of abundant copper throughout the basin. The El Pasoans located a promising copper vein adjacent to Carrasco's mine.[23] They brought men and supplies from El Paso and began to do the assessment work necessary to prove their claim. Francisco Bernál, acting lieutenant governor of New Mexico at this time, petitioned Commandant General Salcedo to relieve him of this office so he could devote his time to mining. Salcedo granted his request, and Bernál worked his El Cobre property for several years.[24]

Bernál's partner Francisco García, a merchant and farmer, was a leading citizen of El Paso del Norte. He raised large herds of cattle and sheep on his Rancho de Santa Teresa land grant seven miles northwest of El Paso. Although Francisco García did not take an active part in working the mines, he handled the supply and transportation for the business.[25] Francisco's brother, Juan Antonio García, was also interested in the El Cobre area. Juan and several others from El Paso brought a flock of sheep to the *Ciénega de San Vicente* (present-day Silver City), about fifteen miles west of the copper mines. He built a stone fortification for protection from the Apaches, and worked some silver mines in the area. Juan Garcia tried to obtain a land grant at the ciénega, but his petition failed because the colony was too small to protect themselves from the Indians.[26]

Carrasco's son, Joseph Ignacio, had more success when he petitioned General Salcedo for permission to take over an abandoned hacienda near Janos. He explained that the discovery of the Sierra del Cobre mines by his father made it necessary to feed many people. The Rancho la Pastoría, abandoned for more than forty years, had enough good grazing land to maintain fifteen hundred heads of cattle and six to eight horse herds. It also contained land on which to plant crops. Violent Apache depredations forced many ranchers and farmers out of business in times past. Salcedo readily approved the plan to reclaim this property.[27]

Once again, José Carrasco needed money. In spite of tax relief and a hacienda, he could not make a profit from his copper mines. He had already depleted much of the native copper easily obtained from shallow pits near the surface. The next phase of development required sinking a deep shaft, more technical skill, and a much larger workforce for the strenuous task of moving earth and breaking rock. This was beyond Carrasco's skill or his pocketbook. In the fall of 1803, he sold his El Cobre copper mine to Francisco Manuel Elguea, a wealthy merchant from Chihuahua City.[28]

After selling the mine to Elguea, Carrasco purchased supplies and put his small labor force to work on his second copper claim. Although this mine produced low grade ore, he hoped to strike a large body of copper as he followed the vein. When this did not happen, he turned to prospecting, hoping to find something that would pay handsomely. He located several claims over the following years, but nothing significant ever developed from them. Because he was often away from El Cobre, some of his claims were taken over by others as abandoned mines. He was frequently involved in litigation trying unsuccessfully to regain what he felt was his. In spite of disappointments, Carrasco never gave up. He continued to make his home at the copper mines and take a proprietary interest in the welfare of the mining camp. One of the last things he did before his death in 1819 was to petition the Commandant General for a deputy with the authority to bring law and order to the camp and a priest to serve their spiritual needs.[29]

Lieutenant Colonel Josè Manuel Carrasco spent twenty years prospecting and mining in the Sierra del Cobre area. Although he failed to make a fortune for himself, his small discovery in Apachería would lead to the later development of the great Santa Rita open-pit mine which became one of the leading copper producers in the United States.

3

Francisco Manuel Elguea—Entrepreneur

Francisco Manuel Elguea could match any modern entrepreneur as an astute businessman. He had wealth, political influence, a thriving mercantile business, and government contracts. He used all of these assets to further the economic development of the Sierra del Cobre mines.

Francisco Elguea was born in Vitoria, Spain. As a young man, he emigrated to Mexico where he established a successful mercantile business in the city of Chihuahua. Elguea took an active part in civic affairs, served on the *Ayuntamiento* (city council) and held various other political offices such as subdelegate of the treasury. He was a lieutenant in the *Dragones Provinciales del Principe*, an elite civil militia corps of Chihuahua.[1] As a member of the prestigious merchant's guild of Chihuahua, Elguea held government contracts to furnish supplies to the frontier presidios. His mule trains brought military supplies, payrolls, tools, food and clothing to the military posts and the villages that grew up around them. Elguea also owned a large ranch called Hacienda del Torreón, located in a fertile valley about thirty-five miles northwest of the city of Chihuahua. Besides grain and cattle, the hacienda raised the horses and mules that Elguea needed to transport goods for his mercantile and supply business.[2]

For several years Francisco Elguea enjoyed profitable business relations with New Mexico. Annual caravans brought sheep, textiles and other items to markets in Chihuahua. At times the government ordered large quantities of woolen blankets from New Mexico to give to the peaceful Apaches "to cover their nakedness." Elguea acted as agent for the New Mexicans, arranging sales and contracts and, with the profits from these sales, providing merchandise for the return trip to New Mexico.[3]

Elguea's ability to negotiate favorable contracts is illustrated by an arrangement he made with the Commandant General in 1803. The city of

Chihuahua needed a new water system. With this in mind, Elguea agreed to furnish supplies to the presidial company at Santa Fe, New Mexico for eight years. In exchange, he pledged his annual profit of six percent to a project designed to bring water into Chihuahua and construct fountains there for public use. He lost nothing by this arrangement because his supply trains returned from Santa Fe with hides, textiles and wool for the Chihuahua market. He also had a contract to sell this uncleaned wool to the woolen mill in San Felipe.[4]

In the latter part of 1803, Elguea acquired Carrasco's copper mine in the Sierra del Cobre. A substantial amount of money and labor were needed to develop the mine into a profitable investment. Labor was no problem for Elguea. Under a system of debt peonage, mine owners could contract for unskilled labor from the prisons by paying off the prisoner's debts. Petty thieves and debtors were required to work in bondage with half of the small wage they earned going to repay the value of the debt. Mine owners could also draft vagabonds and idlers to work in the mines.[5]

By the beginning of 1804, Elguea had a work force of 45 men employed in developing his mining operation. Furnishing food and other supplies for a large number of people in a place far from the nearest source of goods required many resources. Elguea, with an established mercantile business, had no trouble getting the food, clothing, tools and supplies needed at the mine. Ranches in the valley of the Rio Grande furnished sheep, cattle, horses and mules. Farms along the river provided food and grain, and local vineyards produced wine of outstanding quality. Frijol, wheat, maize and cattle came from Janos. Trade with the Valle de San Buenaventura and Sonora's agricultural regions provided food and livestock. Elguea's mule trains carried military supplies to the presidios of Janos and San Buenaventura, where they picked up locally-grown foodstuffs for the copper mines. On the return trip, the mule trains left El Cobre loaded with copper destined for Mexico City.

Mule trains were the lifeblood of the Sierra del Cobre mines. A mule train might have as many as 100 mules, with each animal carrying up to 300 pounds of cargo. A crew of six *arrieros* (muleteers) handled each group of forty to fifty animals. These sturdy beasts covered twelve to fifteen miles a day, negotiating trails too rough for wagons or the *carretas* (two-wheeled carts) typically used in Mexico.

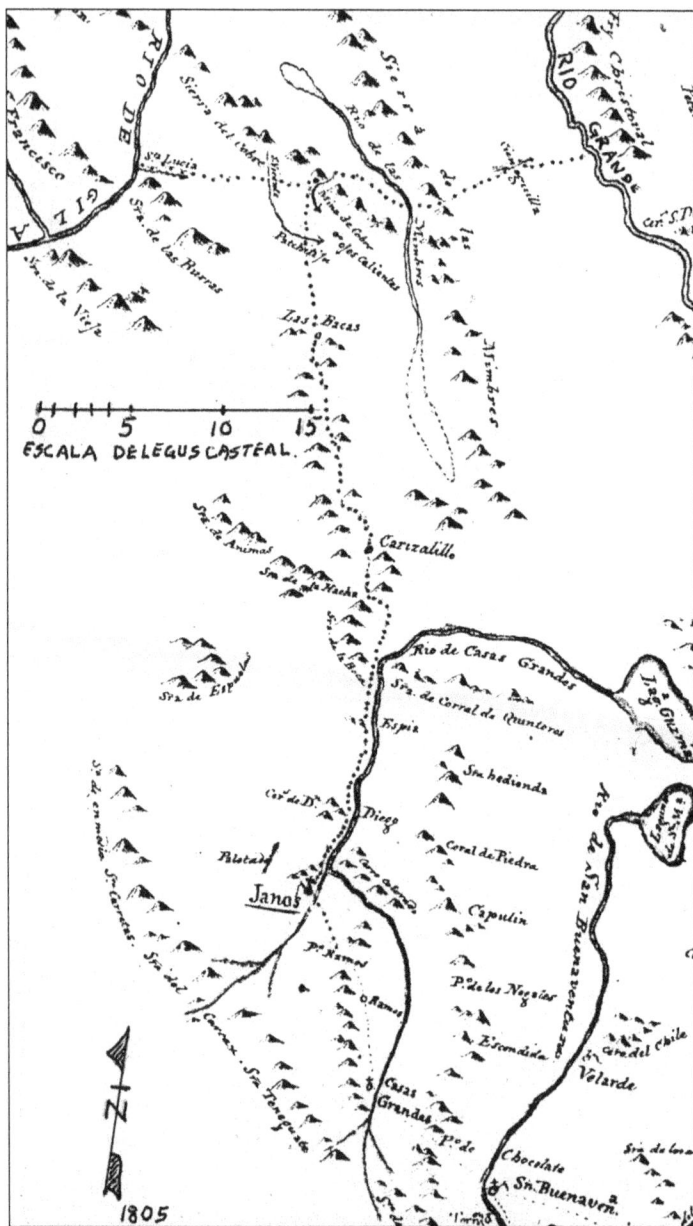

The Trail to Janos. The area of the Santa Rita mine and the trail to Janos was reconnoitered by 2nd Lieutenant Juan Pedro Walker of the Janos Company on order of Commandant General Salcedo in 1805.
Courtesy of Baker, Texas History Center, University of Texas at Austin.

The trail between Santa Rita del Cobre and Janos was long, rough, and dangerous. Apaches were known to ambush unwary travelers along the way. The main trail from the copper mines led southwest through a canyon along present-day Santa Rita Creek to its junction with Whitewater Creek. The trail followed the watercourse for a few miles before cutting across country to the springs at Pachitejú (now Apache Tejo). From Pachitejú the trail went to the next water hole at Ojo de Vaca (Cow Springs). Here it joined the main trail to Carizalillo Springs, continuing southward to Janos on the San Miguel River. The trail passed through Casas Grandes and San Buenaventura and joined the road from El Paso del Norte to Chihuahua City.[6]

Elguea needed enormous quantities of leather goods for his various businesses, especially transportation and mining. Mule trains used leather for packsaddles, bags, straps and other gear which received hard use. The mining industry needed many rawhide bags and sacks to carry the ore, leather buckets to bail water from the shafts, and strips of leather for binding and mending. To meet this need for leather, Elguea arranged to purchase tanned hides from Apaches living at the peace establishments near the presidios. The Indians were allowed to return to Apachería for three or four months during the fall for hunting and gathering of wild plant foods. They were excellent hunters, and often returned with a surplus of animal skins. The Indians received the current price in cash or merchandise, preferring hats, cotton and flannel cloth, and equipment for their horses as items of barter.[7] This resulted in another efficient and profitable operation for Elguea.

With supplies on hand and a work force in place, development of Elguea's mine went forward. Diego Obeso, a native of Castile and a clerk in Elguea's Chihuahua business, was appointed to oversee the mining operation. Carrasco, during his brief tenancy of the mine, had removed the rich ore deposits easily obtained from pits and shallow trenches near the surface. Now the work went underground. There was little planning in the operation or thought for the future. The workers ran a sloping shaft into the hillside to follow the vein in search of the large pockets of high-grade ore. Underground excavation followed the veins and ore shoots through a series of tunnels and galleries branching off in all directions and levels from the main shaft. As the shaft deepened, the workers climbed from one level to another on crude ladders made of stout juniper or pine logs notched for footrests. *Barenadores* (pickmen) extracted the ore using wedges, sledge hammers, heavy iron picks, and long crowbars weighing thirty to forty pounds. They broke the ore from the waste rock and gathered it into *tenatas* (ore bags) of deer or cowhide laced with leather thongs. *Tenateros* (ore carriers) loaded the leather bags on a shoulder pack called a *seroni*, which was held in place with a strap across the forehead, and climbed the notched ladders to the surface. Workers removed waste rock in the same way.[8]

Underground mining. Copper mining methods were primitive in the 1800s. A notched trees served as a ladder to reach the various levels, tallow candles provided light, and workers pried the copper from the matrix with a simple metal bar. Louis Simonin, *La Vie Souterrane.*

Danger of a cave-in made work in the mines a risky business for the laborers. As the Elguea shaft deepened, the workers ran into fractured ground. No more than six feet could be dug without shoring. This was very costly, but necessary for the safety of the workers and the welfare of the mine. Juniper logs, notched to fit overhead beams and bound together with rawhide, supported the walls at weak points in drifts and tunnels. Sometimes pillars of earth, left in place at intervals of six to ten yards, prevented the collapse of the hanging wall. A box-like structure called crib lathing might timber the sides of a shaft and reinforce a mine entrance.[9] Worked-out drifts and abandoned tunnels provided a place to dump the waste rock without carrying it to the surface, and prevented cave-ins.

Elguea sent his first shipment of El Cobre copper to the royal mint of Mexico City in April of 1804. An assay verified the excellent quality of the ore. Native copper, found in combination with red copper (cuprite) and mountain green (malachite), greatly facilitated the refinment of the copper. In the opinion of the administrator of the royal mint, the copper from Sierra del Cobre "is the best copper that has come into this mint in many years." Following this glowing report, the mining commissioner ordered the director of the mint to purchase all the copper produced from Elguea's mine. The suggested price for ore of this quality was twenty-five pesos per hundred-weight.[10]

The Commandant General was pleased with this favorable report. The discovery and development of new mines led to economic growth, territorial expansion and frontier prosperity. Salcedo announced the discovery of the Sierra del Cobre mines to the Ministry of the Indies in May of 1804. He said that the owners had extracted more than 350,000 pounds of copper from the mines since their discovery, and stressed the excellent quality of the copper. In his opinion, "this promises to be a lucrative business that furnishes jobs for the people and benefits the country."[11] Salcido wondered, however, what effect all the increased activity in the midst of Apache lands would have on the fragile peace the Spaniards had worked long and hard to establish with the Indians.

Many Apaches, both peaceful and hostile, lived in Apachería. Some of their favorite campsites were within fifteen to twenty miles of the new mining camp. Salcedo cautioned the miners about the dangers of leaving their camp to prospect in the mountains. Ambush by the enemy could have grave consequences and weaken the settlement before it became well established.[12] He ordered the Janos commander to send troops on punitive expeditions to the mountains and valleys around El Cobre whenever he could spare the soldiers from their regular duties.[13] A small squadron of soldiers patrolled the route to El Cobre and protected those traveling to and from the mines. In addition to the economic value of the new mines, El Cobre was situated in a strategic location from which to observe the activities of the hostile Indians. A permanent settlement in the middle of Apachería could be a military advantage in dealing with the fractious Apaches.

El Cobre copper continued to pour into the treasury warehouse in Mexico City. In the first four months of 1805, the Santa Rita mines turned out 96,200 pounds of the rich red metal. Production from May through August resulted in 122,737 pounds of copper, and between October and December the yield rose to 137,235 pounds. The administrator of the district reported that the depth of the veins demanded more underground work before extracting a greater amount of ore. However, the quality of the copper was so superior that it did not require smelting before it went to Chihuahua. Each shipment of ore contained as much as twenty-one granos (168 ounces) of gold; all signs indicated that the mines would continue to yield abundant copper.[14]

Expectations for the Sierra del Cobre mines soared that summer when prospectors found evidence of gold in the streams and side canyons four or five miles from the copper mines. Soon, men were working the sand and gravel of the watercourses with shallow wooden bowls, finding flakes of gold and sometimes a sizable nugget or two. As others arrived to seek this precious metal, a rustic camp of brush shelters sprang up around the placers. They called the settlement *Real de Santísima Trinidad del Oro* after Carrasco's original gold claim of 1801.[15]

A search for the source of the stream deposits led Blas Calvo y Muro to a vein of gold in a hillside near the camp. He registered the claim on July 5, 1804, naming it *Santa Rita de Casia,* honoring the saint of the impossible and the advocate of desperate cases.[16] Blas Calvo was a merchant from Chihuahua and an associate of Francisco Elguea. Calvo came to the copper mines soon after Elguea took over Carrasco's mine. He located a vein of copper and set about developing it, but failed to hit the high-grade ore that the Elguea mine produced. The two men formed a partnership to work the Santa Rita gold mine, with Elguea furnishing the labor force and Calvo overseeing the mining operation. A shaft was sunk along the vein, and within a few months they had taken out gold worth three thousand pesos, and had another six hundred pounds of ore on hand.[17]

Dreams of wealth lured others to the Sierra del Cobre. Prospecting for gold continued along present-day Whitewater Creek, up Gold Gulch, and in an arroyo called *La Ciéneguita*. By the end of the year prospectors had discovered sixteen veins of gold. Following an inspection of the new area, the Janos commander reported that in his opinion, "a hundred miners could establish claims here." He concluded, "Only encouragement is needed for the people of Chihuahua, Parral, and Arispe to take advantage of these discoveries. They have all the signs of being one of the most important of the province."[18]

The excitement of the gold discovery proved to be of short duration. Little rain fell in 1805; springs dried up and lively streams shrunk to a trickle. Without a good supply of water, miners could not work the placers or recover gold from the vein ore. After producing some 170 marks (1360 ounces) of gold, the output decreased sharply.[19] The drought continued into the next year. The streams became dry arroyos, and work at the placers came to a standstill. El Paso miners Francisco Bernál and Francisco Meras, undaunted by the arid conditions, registered three additional claims located in the El Cobre area in July 1806.

The copper mines continued to yield large amounts of copper: 140,838 pounds produced between January and May 1806, followed by 160,776 pounds during the next three months. As the shafts deepened, the ore grade became ever richer. There appeared to be no end to this abundant deposit of the rich native metal.[21]

Francisco Manuel de Elguea died at his home in Chihuahua on September 18, 1806, before he fully realized the success of his mining investment. He left a wife, Doña María Antonia Medina, two young sons, Mariano and Francisco, and two daughters from a previous marriage. Doña María became principal executor and guardian of the estate. Elguea's agent and general attorney, Francisco Jáurequi, and his friend Captain Salvador

Uranga were designated coexecutors. Jáurequi took over the administration of the Elguea estate's business affairs. Under his efficient guidance, the El Cobre copper mines continued to prosper.[22]

The entrepreneurial skills of Francisco Manuel Elguea firmly established the economic value of the El Cobre property. His interest in the development of the Sierra del Cobre mines gave the business a reasonable chance of success. He had the resources to invest in improving the mines and expanding his holdings. When he died, Elguea owned the principal copper mine plus several other copper claims and all of Blas Calvo's interest in the Santa Rita gold mine. Under Elguea's leadership, production of copper increased from now-and-then shipments to long mule trains, loaded with copper, making the trip to Chihuahua every three months. This success attracted other miners and prospectors to the area, and the camp began to take on a semblance of stability. The larger and stronger the camp, the better the people could withstand Apache aggression.

4

Real del Cobre

The Sierra del Cobre mining camp in the heart of Apachería was difficult to colonize, hard to supply, and vulnerable to attack from hostile Apache bands who roamed the area. The Real del Cobre was completely isolated from other frontier communities. El Paso del Norte (now Juarez), the nearest settlement to the east, was 150 miles away; the presidio of Tucson, in Sonora, was about the same distance to the west. Northward high mountain ranges separated El Cobre from the towns and pueblos of New Mexico. The commander of the presidio of Janos, 150 miles to the south, had jurisdiction over the copper mines. Only one trail, built over rough and dangerous country, led to the new mines. Apaches often lay in wait along the trail, hoping to ambush unwary travelers or supply trains. Santa Rita del Cobre survived and thrived under these adverse conditions because of the fine quality and abundance of the copper, and the strong support from the Commandant General.

During the early years, when Carrasco struggled to make his copper mine a paying proposition, little thought was given to establishing a permanent camp. His workers, probably not more than ten men, lived in rustic brush shelters around the claim. They refused to remain at the mine when Carrasco took his ore to market because they feared attack from the Apaches who roamed freely in the area.[1]

After Elguea took over Carrasco's mine in the fall of 1803, he brought in a colony of forty-five workers to form the nucleus of a permanent settlement. Some of the men brought their families with them. Winter was fast approaching, and there was an immediate need for shelter. Soon, a jumble of hastily erected brush huts clustered in small groups around the mine. As time allowed, more permanent dwellings, called *jacales,* were erected. This was a simple one-room hut made of upright poles fastened together and chinked with adobe. It had a thatched roof, a hard-packed dirt floor, and animal skins to cover the door. There was little furniture other than bed rolls, covered with a blanket to make a seating area during the day.[2]

Workers eventually erected a large, flat-roofed adobe building to house the mine superintendent's quarters and office. The building had a storage room for supplies and likely also a general store, which furnished necessities for the workers. Breastworks and a fortified *torreón* (tower) provided a refuge where the people gathered in case of attack from raiding Apaches.[3] Buildings were scattered at random intervals around the mines. Captain José Tovar, commander at Janos, recommended a more orderly arrangement of the camp for better protection. He suggested, "the town can be formed in two parts along either side of the arroyo (Santa Rita Creek) that flows through the camp. There is room to build eighty medium size houses, leaving a small plaza in the middle."[4] Although Captain Tover's suggestion made sense, there is no indication that the residents of El Cobre followed the good advice of the captain. Instead of a strong, centralized community, the settlements continued to gather around each respective mine.

The gold mines at Santisima Trinidad del Oro, a few miles downstream from the copper mines, were especially vulnerable to Apache incursions. The camp was a cluster of brush shelters and crude huts around the placer workings. Captain Tovar advised the gold miners to erect a stockade and a sturdy building that would offer some protection against the Apaches. The sentries were instructed to be especially vigilant because the area around both camps was covered with an oak and pine forest and visibility was limited.[5]

Ranching and farming went hand-in-glove with the opening of new mining districts. Animal products were essential to the business, not only for meat to feed the workers, but also for by-products such as leather and tallow. Tallow candles were the only means of illumination in the mines. To meet these needs, Elguea brought in a herd of cattle and a flock of 500 sheep, with stockmen and a shepherd to tend them. *Pachetejú* (now known as Apache Tejo), a few miles south of the placers, met all the requirements of a good stock ranch. There was plenty of pasture land where grama grass grew high as a horse's belly. A spring-fed stream provided water for the area. Herdsmen, armed with rifles, took the large herds of cattle, horses and mules to graze there, even though the animals were vulnerable to raiding Apaches when away from the settlement.[6]

A successful mining operation required large amounts of foodstuffs and grain. The land around the mines was largely useless for farming, except for an orchard and small vegetable plots. The Janos commander suggested that the colonists open a farm in the fertile Mimbres River valley located eight miles to the east. Farmers could raise wheat by irrigation and corn or maize through dry farming. Pachetejú, also suitable for farming, would accommodate thirty-five or more acres of irrigated crops and at least a hundred acres of dry farming.[7] This would do much to relieve the burden of hauling grain all the way from Janos or San Buenaventura. Tovar also thought the miners could clear land between the copper camp and the placers and have many acres for planting corn, wheat and frijol.[8] The El Cobre colonists tried to carry out this excellent plan to provide a local source of grain and food. Unfortunately, farming was never very successful because marauding Apaches helped themselves to the crops, and then destroyed the fields.

With all the growing traffic on the trail to El Cobre, and the influx of people coming into the area, it is surprising that the Apaches had not yet raided the settlement itself. Tracks nearby and smoke signals on the mountains indicated that the Indians were aware of everything going on in Apachería. Apaches living at peace establishments near the presidios were allowed to return to their homeland in the fall to hunt and gather wild plant foods. It was particularly stressful for El Cobre citizens when Apaches assembled in large numbers in the neighboring hills. For the Indians, it was a time of reunion with relatives and friends still living in Apachería; for the miners, it was terrifying.

The new settlement was poorly equipped to deal with Indian raids. Primitive brush shelters and thatched-roofed huts offered little protection against Apache arrows. The workers at the copper mines had various kinds of weapons, although most of the peons lacked the skill to repel raiders that operated with the notorious swiftness of the Apaches. Of forty-five men employed at the Elguea mines, five owned muskets, three fought with bows and arrows and the remainder used slingshots, clubs or whatever they might fashion into a weapon. Five of Carrasco's men carried guns and five had lances or bows and arrows. All the men went to work with their weapons at hand and there were sentinels posted to guard against a surprise attack. Captain Tovar ordered a reconnaissance troop from Janos presidio to patrol the area around the mines during the hunting season which passed without incident.[8]

The Sierra del Cobre mines were of particular interest to Commandant General Salcedo. With the high quality of the copper coming from the mines, and the discovery of gold in the area, El Cobre showed excellent prospects of becoming an important new mining district. The trail from Janos to El Cobre saw much activity as more miners and prospectors came to seek their fortunes. Mule trains brought supplies to the camp and returned to Chihuahua loaded with copper. The Commandant General did all he could to insure the safety and well-being of those involved in this new colony. He ordered the Janos commander to send troops to escort the mule trains and people traveling the long and dangerous trail. The soldiers monitored the activities of the Apaches, and warned them to stay away from the new settlement.[9] The mining camp served as an outpost with a military detachment of ten to twenty soldiers stationed there from 1804 to 1809. They guarded the copper camp and the placers, and patroled the surrounding area. Additional troops were sent in at the first sign of trouble.[10] The presence of soldiers at the settlements was probably the only thing that kept the Apaches from driving Spanish colonists out of Apachería in these first few years.

Mining camps of the frontier were notoriously rough and disorderly. Drinking and gambling, although prohibited, remained favorite pastimes, often leading to disobedience and violence. Prospectors jumped claims and abandoned mines as quickly as they opened them, often neglecting to register their workings. Rules and regulations were ignored or had to be modified to meet existing conditions. Adding to the problem, the Sierra del Cobre mines, although in the province of New Mexico, operated under the jurisdiction of the military commander at Janos, Chihuahua, 150 miles to the south. Communication was a serious problem. If El Cobre had trouble with Indian raiders, a mining dispute, or any crime of a serious nature, a message was sent to the Janos commander. It might take the messenger as much as a week to reach Janos unless he ran into an Apache ambush along the way. If the problem required orders from the Commandant General, it took another ten days to get an answer from Chihuahua, and another week to notify El Cobre of the decision or rush to their assistance.

Salcedo decided he needed to take steps to insure law and order in the new camp. In August 1804 he appointed Diego Obeso, superintendent of the Elguea mines, as *comisario de justicia* (deputy justice). The Elguea mining enterprise dominated the camp. The Elguea mine was the best developed and most productive mine in the area. Elguea's mule trains took out copper and brought in supplies; they were the area's link to the outside world. The superintendent of the Elguea mines always served as administrator of justice and more or less controlled the camp. The new official heard petitions, investigated crimes, punished the transgressors, and practiced judicial proceedings under the direction of the Janos commander. He sent monthly reports to the Commandant General on the condition of the mines and the number of people living there. Salcedo was especially interested in the activities of Apaches in the area and various measures taken for the defense of the settlement.[11]

New Mexico had few active mines except those at the Sierra del Cobre and had no Territorial Deputación de Minería at this time. Miners from the El Cobre area had to travel 360 miles to Arispe, in Sonora, or 460 miles to Chihuahua City to file applications and register claims. Salcedo notified the Deputación de Minería in Chihuahua that "mines already discovered and those discovered in New Mexico in the future will be under the Real del Cobre mining district, which has now been established." This was the first mining district in New Mexico. Justice Obeso received authority to register mining claims and placers and to take care of all mining matters according to the Royal Mining Ordinances of 1783. New Mexicans did not take advantage of Salcedo's directive. Santa Feans found it impractical to trek across the mountains to El Cobre to register their mines, so they continued to register claims with the New Mexico governor. It was a great help, however, to those conducting business at the Sierra del Cobre mines.[12]

The official name, *Real de Santa Rita del Cobre*, now occasionally appears in records pertaining to these copper mines. Saint Rita was the saint of the impossible and the advocate of desperate cases. It was a particularly appropriate name for a mine operating successfully under adverse conditions.

The abbreviated form, El Cobre, continued to appear in reports and orders. The more familiar *Santa Rita* will be used for the remainder of this story to avoid confusion.[13]

Daily life was dangerous, dreary, and monotonous at Santa Rita. The work was hard and the hours were long (twelve hour shifts), which left little time for leisurely pursuits. Religious holidays were observed, however, and an occasional fiesta was held. Gambling and drinking were prohibited in the mining camp, but often indulged in furtively. Bartering was among the few spare time pleasures available. The miners had little money to spare, so they often traded their meager possessions for other items, frequently food. The case of the missing sheep is a good example of a bartering operation that went sour.

It all started when fifteen sheep disappeared from a herd of five hundred belonging to the Elguea *hacienda de minas* (mining establishment). The justice asked Julio, the young shepherd in charge of the sheep, to explain the shortage. It appeared that Julio either sold the sheep, gambled them away, or devoured them with the help of his friends. In questioning the accessories to the crime, one man confessed that he traded Julio a cloth cigar case for half of a sheep carcass; another paid twelve reales for a whole sheep. Juan, a lad of sixteen who worked as an ore carrier, traded his sombrero for mutton, while another friend admitted winning an animal in a game of chance. He knew gambling was illegal, but he occasionally did so to divert his mind from the possibility of being killed by a cave-in at the mine. Francisco, a horse guard for the Elguea mining hacienda, gave Julio several cartridges and some paper in exchange for a sheep. Obviously, all were guilty to some degree, as they knew the sheep were stolen.

The judge appointed guardians to see to the welfare of the two young boys, Julio and Juan, who evidently had no family to care for them. They were sentenced to work as water bailers, one of the toughest jobs in the mines. Due to the shortage of experienced miners, the supervisors of the other offenders paid their fines so as to avoid jail sentences and get the men back at work.[14]

Many of the mine's employees had their families with them. Santa Rita del Cobre had no church, but the Elguea estate paid anywhere from 150 to 200 pesos for a priest from either Janos or El Paso del Norte to visit the mining camp periodically and administer to the spiritual needs of the inhabitants. After Elguea's death, his estate continued to provide this service. From the report of Father José Antonio Ulibarrí, priest at the Church of Nuestra Señora de Guadalupe at El Paso, it appears that the mining camp was thriving. On August 15, 1806, Father Ulibarrí journeyed to the mining district where he baptized eight children born in the settlement between December of 1805 and August of 1806. In the summer of the following year, Father Augustín Gómez Platón, chaplain of the Janos presidio, came to hear confessions and minister to the faithfiul.[15]

Six years later, Juan Tomas Terrazas. the priest from El Paso del Norte, expressed concern for the spiritual welfare of the people at Santa Rita. He reported that there were more than four hundred souls living in the area and still no church or priest to serve them. There was "only a small piece of consecrated ground, under a juniper tree, designated for the bodies of the people who died there."[16] Between visits from the priests, people traveled to either Janos or El Paso for confessions, according to their preference; for baptisms and marriages they usually went to Janos because the road was better in that direction. There is no evidence of a chapel, church, or resident priest at Santa Rita during this period.

The Indians of Apachería continued to raid settlements of Sonora and Chihuahua, but hostilities did not disrupt the Santa Rita operation to any great extent until the summer of 1807. In their search for plunder, the Apaches carried out a destructive raid in Sonora on the Tumacacori mission (20 miles north of the present U.S.-Mexican border). They returned to Apachería with much loot, including the entire mission's herd of horses. A Sonoran military unit and troops from Chihuahua's frontier presidios joined forces for a campaign against the Indians. Because of the favorable location of Santa Rita del Cobre in the midst of Apacheria, the camp served as a base of operation for the military. A number of stolen horses were repossessed and held at the mining camp until they could be returned to their rightful owners.[17]

Santa Rita's temporary role as a military outpost led the Apaches to seek revenge. On June 25, 150 Apaches conducted their first raid on Santa Rita. They killed a soldier, wounded a guard, and drove off a herd of cattle. Troops from Janos were sent to protect the miners.[18] Salcedo reminded the residents that it was the responsibility of the people who work there to bear arms for their defense and the safety of their possessions.[19] The inhabitants lived in a constant state of uncertainty, not knowing when the Apaches would pay them another visit.

The following year the Indians struck again. This time they attacked a squad of troops who had gone out from the mining camp to cut forage for their horses. One soldier was killed and three were wounded; their weapons, two horses, and two mules were stolen. In the next incident of record, six Apaches raided the livestock pastured some quarter of a mile from the camp. They killed one guard and were driving off the herd when the other guard fired his gun. The noise brought out the detachment of troops and several armed peons, who drove off the enemy and managed to round up most of the animals.[20]

Life was hard at Santa Rita. Work in the mines was arduous and dangerous. There were few diversions and only the bare necessities of life to sustain the inhabitants. Although the Indians raided the outlying areas of the settlement now and then, they only inflicted minor damage, and the mines continued to prosper. This extremely rich property produced over one and a half million pounds of native copper in the first six years of operation. It was now ready to face any changes the future might bring.

5

Mines, Metallurgy and Markets

Santa Rita del Cobre now entered a crucial period with the future of the copper mines in jeopardy. Government regulations required a new and more costly method for treating the copper before sending it to the royal treasury. In 1809 Spain ended its control over the purchase and distribution of Mexican copper, leaving those involved in the copper business to find their own markets. With copper no longer under government control, the price dropped. Commandant General Salcedo withdrew military support from the isolated mining camp, leaving it vulnerable to attack from the Apaches. Could the fledgling Santa Rita del Cobre find a way to overcome these major setbacks and continue to prosper?

In the early years, prospectors came to Santa Rita with great expectations. They staked out claims and harvested the native copper from weathered veins near the surface. When the surface ore played out, few prospectors had the resources for further development of the claim. They moved to another promising outcrop and started over, leaving the valley littered with test pits and trenches.[1] Underground miners had little knowledge of geology and lacked the technical skills to plan excavations. They tunneled into the hillside and sank a vertical shaft along the weathered vein. Rich ore pockets occurred along a single vein or where two veins intersected. The simple goal was to follow the course of the vein in hope of striking a large pocket of high-grade copper. Without systematic designs or projections for the future, the mine became a warren of low, twisting tunnels and narrow drifts at various heights. As the shaft went deeper, this haphazard arrangement resulted in difficulties in extraction, drainage and ventilation.

In 1805 Commandant General Salcedo reported to Spain that the copper produced at Santa Rita was of such superior quality that it did not need smelting. The ore also contained enough gold to pay the high cost of transporting the copper to the royal treasury in Mexico City. Every four months between five and six thousand pounds of copper went south on the mule trains. Elguea's mine was by far the best in the district and the copper showed no sign of diminishing, despite the careless mining methods used for extracting the ore.[2]

Word of these fabulous copper mines reached the United States by way of the journal of Lieutenant Zebulon Montgomery Pike. In 1806 the United States government sent Pike to explore the boundary between the Louisiana Purchase and New Spain. Lieutenant Pike and his men strayed further west than anticipated and entered New Mexico territory in 1807. Spanish authorities took the Americans into custody for trespassing, and escorted them to Chihuahua to appear before Commandant General Salcedo. Pike gathered much valuable information on his trip down the Rio Grande valley. At El Paso del Norte, he stayed at the home of Don Francisco García, a well-to-do merchant and farmer. Garcia once owned a mining claim at Santa Rita with Francisco Bernál. It is likely he gave Pike a glowing report on the new copper mines.[3] While Salcedo detained him in Chihuahua, Pike also had the opportunity to discuss the Santa Rita mines with the Commandant General.

In his journal, published in 1810, Pike stated that the only active mine in New Mexico was a copper mine in a mountain west of the Rio del Norte. He wrote: "The mine produces 20,000 mule-loads (7,000,000 pounds) of copper annually. It also furnishes that article for the manufactories of nearly all the internal provinces."[4] Obviously Pike was referring to the Sierra del Cobre mines, although his production figures are questionable. According to Salcedo's report to Spain for this period, the copper mines produced close to 20,000 *arrobas* (507,400 pounds), not 20,000 mule-loads, of copper in the year preceding Pike's visit. Considering the primitive methods used in mining the ore, and the hazardous conditions under which the miners worked, this adds up to a good production record.[5]

The copper came from the mine in all sizes and shapes, from large nuggets to sheets and stringers. It was so pure that it could go to Mexico City in its natural state. This made an awkward load for the mules, even when packed in wool and loaded in rawhide bags.[6] The solution was to reduce the ore to a uniform size and shape. The first smelting at Santa Rita took place in small open pits lined with a mixture of clay and ash. Laborers broke up the ore by hammer and placed it in the pit. Heat provided by charcoal, and the blast from a hand-operated bellows, reduced the ore to a molten state. Workers skimmed off the slag and the remaining metal hardened in the pit. This primitive method, used in other copper mines of New Spain, resulted in a crude ingot called a *plancha*. It took about eighty hours to smelt eight hundred pounds of ore.[7] Planchas were much easier to pack on mules than rough pieces of native copper.

In October 1807 the royal treasury ordered copper submitted as bars of a standard weight and fineness rather than crude planchas.[8] The new smelting process required the use of a furnace similar to the *horno castellano* used in the refining of silver. Built of adobe, it stood eight feet tall and eighteen inches in diameter. The metallurgical process was simple. Workers loaded the furnace with charcoal through the open top. First they fired it to preheat the furnace and hearth, lessening the prospect of damage to the furnace from the intense heat to follow. When the furnace reached the proper stage, workers threw in the ore mixed with flux. Flux was composed of material that melted easily and helped to heat the ore and absorb the impurities from the metal. More charcoal was then added. Hand-operated bellows produced a blast of air through holes in the back of the furnace, thereby increasing the heat. As the ore and flux melted, it flowed from the furnace onto the hearth, which sloped to allow the molten mixture to run into a forehearth. Flux combined with the gangue, or waste portion of the ore, produced slag. This waste, lighter than the metal, rose to the top, where a worker removed it with a hooked bar or fork. He then ladled the liquid copper into heated molds where it cooled to form bars. The workers recharged the furnace and repeated the procedure on the next batch of ore. It took about eight hours to treat a run of good ore. If they encountered no trouble, the smelting continued for three or four days and nights.[10]

Molds for casting the copper were typically made of metal, but stone molds were also used. Among artifacts found in the Santa Rita area was a set of four molds cut side by side in a large rock. Each recess was six inches by twenty inches, and four and one-half inches deep. Two ingots of one hundred fifty pounds each would make a good mule load. The treasury paid twenty-four pesos four reales per hundred pounds for copper in bars, with the peso roughly the equivalent of one dollar at the time.[11]

Old Adobe Furnace. This is the type of furnace used to refine copper ore at Santa Rita del Cobre in the early 1800s. Fayette Jones, *New Mexico Mines and Minerals*.

A *hacienda de benificio* (establishment for reducing the ores) usually contained a roofed area housing the furnaces and sheds for storing charcoal and ore. The hacienda might also include quarters for the workers and their families, the house of the owner or administrator, and a company store and office. A high wall surrounded the entire area, providing a corral for horses and mules when Apaches raided the camp.

The introduction of furnaces at Santa Rita created a demand for large amounts of charcoal, as the reduction of ore required a plentiful supply of fuel. Charcoal produced twice the heat given off by an equal weight of dry wood, and was well suited for use in smelting oxidized ores. Mountains surrounding the mines were heavily timbered with oak and juniper, both good materials for making charcoal. Mesquite also produced good charcoal, but pine was too soft. The *carboñeros*, who made the charcoal, set up their camps in the forest, moving from place to place as they depleted the trees. They cut the wood into pieces, two to three feet long, and stacked it in piles several feet high in a cleared area. The carboñeros then solidly packed clay over the surface of the pile to close out the air, leaving an open hole for firing. They closed the hole after firing and allowed the pile to smolder for several days until all the wood carbonized and was ready to use in the furnace.[12]

Salcedo's report to Spain on November 1, 1808, shows the success of the new smelting process. "The copper mines in Sierra del Cobre, between the provinces of Nueva Vizcaya and New Mexico," he wrote, "have experienced some decrease in their yield, but have still produced (March-August) 120,276 pounds of smelted copper and 24,051 pounds of ore." Placing the mines between the provinces of Nueva Vizcaya and New Mexico suggests that the authorities still considered the Sierra del Cobre in no-man's land.[13]

The next major change at Santa Rita occurred when Spain gave up her monopoly on the copper market. Spain controlled the distribution of Mexican copper beginning in 1799, when King Charles IV decreed that the royal treasury would purchase all the copper produced in New Spain at a fair price. Between 1803 and 1809, the treasury bought 8,312,250 pounds of copper valued at close to two million pesos.[14] More than 1.5 million pounds of this copper came from the Santa Rita del Cobre mines. The mining and refining of copper ore increased to such an extent during this time that copper filled the government warehouses. Napoleonic wars raged in Europe, Napoleon invaded Spain, and thus no ships were able to transport the copper to the motherland. Early in 1809, the viceroy declared the end of the government's copper monopoly. The treasury would accept copper already purchased and now en route to the royal warehouse for two months. After that time, they would accept no more.[15] Santa Rita del Cobre produced 111,420 pounds of smelted copper in the six-month period before the government monopoly ended. The mines had on hand an additional 43,039 pounds of ore for which they needed to find a market.[16]

With the end of government control, large amounts of copper became available on the open market. There was fierce competition for buyers, and so the price dropped sharply. Commandant General Salcedo expressed doubts concerning the future of the Santa Rita mines. Troops from the presidios could not provide protection for the mines because they were needed elsewhere to quell insurrections springing up across Mexico. Salcedo notified the Santa Rita administrator that he could no longer support this mining operation. Now, the miners themselves had to furnish their own defense, assure their own safety, and provide armed escorts for the mule trains.[17] Fortunately, copper was in great demand. Spain's ten-year copper monopoly created a domestic shortage of that useful metal. Mexico's armament factories needed copper for cannons. The mint added copper to coins of silver and gold to make them more durable. They used bronze (an alloy of copper and tin) instruments to check the accuracy of weights and measures. The sugar industry required copper vats of various sizes to reduce the cane juices, and copper cylinder casings and gear parts for the crushing mills. Coppersmiths produced a wide range of vessels, kettles, and utensils for business and domestic use. Churches used copper and bronze for fittings, bells, statuary and altar pieces. Metal workers and artisans, who crafted items for commercial and domestic use, suffered from a scarcity of copper during the Crown monopoly. They could buy small amounts from the royal treasury or purchase copper of poor quality on the black market, but the price was overly high. Many were forced to close their shops. With the end of the monoply, copper was now available to meet the demand.[18]

The Elguea family may have established a coppersmithing business near their home in Chihuahua City at this time. A later census lists this *cobreria* (copper shop) as having twelve employees. Ten coppersmiths also worked at the Elguea's Hacienda del Torreón.[19] Santa Rita del Cobre also manufactured copper goods. Janos chaplain Augustín Platón paid his annual visit to Santa Rita in 1809 to minister to the miners and their families. He also placed an order for three copper bells for a new church he was building in Janos. Father Platón died, however, before he could pay the bill of one hundred eighty pesos for the new bells.[20]

Another copper bell, said to have been forged in Santa Rita del Cobre at an early date, eventually found its way to St. Louis, Missouri. Journalist William Dean Howells saw this bell in a St. Louis warehouse in 1858. He described it as "an old Spanish Bell, lately brought from Santa Rita in New Mexico, bearing the date 'A.D. '808, (1808) and some illegible legend in Latin. It was cracked, and bruised and bent. The metal seemed almost entirely copper--perhaps sweetened according to the superstition of the old founders, with silver coins and jewels."[21]

Copper Bells. These ancient copper bells, from an old church in Janos, are said to be made from Santa Rita copper. Humble collection.

Mexico's northern frontier suffered a severe drought in 1809. Streams became dry arroyos and rivers ran shallow. Both mining and agriculture suffered from arid conditions. A shortage of corn and grain sent prices soaring to four times their normal rate. Placer mining declined throughout the frontier.[22] At Santa Rita del Cobre, lack of water shut down the placers, but the Elguea copper mines continued producing an abundant amount of ore in spite of the drought.[23]

Santa Rita met every challenge to the copper mining industry with a viable solution, and the mines continued to prosper. The copper was plentiful and easy to separate from the waste rock. Its purity made it easy to refine, even with the crude furnaces then in use. Technical problems sometimes slowed production; fluctuating markets and falling prices caused anxiety. The miners lived with the constant threat of Apaches raids. Nothing could stop the amazing outflow of the rich native copper from Santa Rita del Cobre.

6

José Baca – A Successful Tenure

Francisco Jáurequi, executor of the Elguea estate, needed a new manager for the Santa Rita del Cobre mines. Santa Rita had four mine superintendents in six years; none had remained on the job for more than a year or two. Jáurequi wanted someone dependable, who knew the mining and smelting business. The manager, as deputy justice, must also maintain law and order in the rowdy mining camp. José Baca qualified for the job. He directed Santa Rita through eleven very difficult years in which Mexico struggled to win independence from Spain.[1]

José Baca took charge of the Elguea mines in the spring of 1810. Under his direction, the mining and smelting operation thrived. The Elguea estate now owned most of the mining claims in the Santa Rita del Cobre area. Some claims were acquired by purchase and others by reclaiming abandoned mines. It was more economical to consolidate all the claims under one company to justify the great expense involved in mining. If a claim was not worked for four months by at least four men, it was considered abandoned. To avoid losing any claims for lack of workers, José Baca allowed other miners to work them under a lease. Part of the profit went to the Elguea estate as the lease fee.[2] Carrasco's early discovery, sold to Elguea in 1804, was by far the most productive of the Santa Rita mines. The underground work went deeper, and native copper appeared in large pockets and great room-size masses. What good mine administrator could ask for more?

One of Baca's primary duties as deputy justice was settling disputes over mining claims. His decisions usually favored the Elguea interests. One example is the case of Francisco Chavez, who worked for the company as a pickman, timberman, and general miner. He left the service of the Elguea mines to work his own claim. Chavez stated that he had difficulty leaving the firm because he owed the company more than two hundred pesos at the time. This suggests that they hired him under the debt peonage system. The company took over a debt owed by Chavez, who was now obligated to work at the mine until he paid off this debt. José Baca was generous. He gave Chavez his freedom for only one hundred pesos, which Chavez agreed to pay with four loads of copper.

Chavez ran into trouble, however, when he filed a claim on a copper mine in a small canyon near the principal Elguea mine. Chavez stated that the mine had not been worked for sixteen months. It was in a state of ruin with tunnels and shafts caved in and pillars broken due to the neglect of the workers and supervisor. Justice Baca denied the claim, stating that it belonged to the Elgueas and was not an abandoned the mine. It had only been idle for a short time due to the shortage of tools and laborers who were working in other mines. Chavez felt that he had not received an impartial decision from Baca, and took his case before the Deputatión de Minería. Francisco Jáurequi, agent for the Elguea estate, verified Baca's testimony and Chavez lost his case.[3]

With the copper mines producing abundantly, Baca could focus on expanding the company holdings. Outcrops of copper covered the floor of the Santa Rita valley. Baca selected one that looked promising and put a crew of laborers to work digging a test pit to follow the vein. Meanwhile, José Carrasco returned to Santa Rita from one of his many prospecting trips. He filed charges against Baca claiming that he (Carrasco) had already started an excavation not fifty feet away. He explained that due to a shortage of labor, he was unable to hire workers. According to Baca, his men had worked on the new pit for several months. He refused to give up the claim, stating that he was registering the mine for the Elguea estate. They called the new discovery *San José* after the patron saint of miners. The San José mine became one of the richest copper mines in the area. The mine was abandoned and rediscovered several times during its long and productive history.[4]

Gold continued to attract prospectors and miners to the Santa Rita area. A few hardy souls earned a living by panning the stream beds for flakes and nuggets until a severe drought made gold panning unproductive. In the spring of 1810, Janos resident Diego Gonzales Rueda arrived at Santa Rita with a party of thirty-eight men eager to "strike it rich."[5] Apaches had recently raided the camp which made it extremely dangerous for small parties of prospectors to roam the surrounding hills. Rueda and his men made their headquarters at Santa Rita and went out to search the area of Nogales Canyon (later Hanover Creek) for new placers. They stayed in a group within four or five miles of the settlement, and were careful to post a guard while they prospected. Rueda must have had some success, because he and his wife listed Santa Rita as their place of residence in 1836, when they served as godparents at the baptism of James Kirker's son José.[6]

Many frontier mining camps earned reputations as rough and disorderly. The miners worked long hours at hard labor in small underground tunnels lit only by tallow candles. In their free time they turned to gambling and drinking for entertainment. Both pastimes were against camp rules because overindulgence usually led to brawls. Justice Baca had a wry sense of humor and his own unique way of handling an infraction of the rules. When he found Simón Molinares in possession of a keg of brandy, he simply confiscated it and emptied the contents of the barrel onto the ground. This action so incensed Molinares that he filed charges against Baca protesting his method of enforcing the law. Salcedo, upon learning of the matter, commended Baca for carrying out his orders, and suggested that he follow the same procedure whenever necessary.[7]

One of Baca's most difficult tasks was dealing with the Apaches. He was responsible for the safety and welfare of some four hundred people now living at Santa Rita. A settlement of this size was considered a thriving community on the sparsely populated frontier. As production at the mines increased, so did the proficiency of the Apaches. They could move into a herd of horses and mules and have the animals on the run in a short time. If the Indians succeeded in taking the entire herd, business at the mines came to a standstill until the manager could replace the animals. A mule cost between twenty-five and thirty-five dollars on the frontier. Losing an entire herd to the Apaches was a serious loss to the mines.

Mexico's fight for independence from Spain drew troops and money from the northern frontier. Budget cuts forced Commandant General Salcedo to suspend rations to the Apaches living at the peace establishments near the presidios. Salcedo ordered a plot of ground assigned to each ranchería. The government issued seeds and tools, and encouraged the Indians to raise food to support themselves. Salcedo's plan caused much resentment among the Apaches. They accused the Spaniards of breaking the terms of the peace treaties which promised rations for the warriors and their families. Many once-peaceful bands left the reservations and returned to Apachería. They became self-sufficient by raiding the settlements and ranches for their rations. Again the frontier provinces were subjected to the widespread devastation of plundering Apaches.[8]

Salcedo launched a military campaign against the marauding bands, followed by a tentative peace agreement. He restored the rations on a reduced scale and assigned the Apaches to specific areas of Apachería. Mogollón bands would establish their rancherías in the area from the Sierra del Cobre west to the Sierra Negrita, including the Mogollón and Mimbres mountains. He assigned the area from Janos, San Buenaventura and Carrizal north to the Sierra de las Burras and Ojo de Santa Lucía (Mangas Springs) to Indians from the presidio peace establishments.[9] The treaty was a tenuous affair, held together only by the restored ration system. With Apaches in every direction, it appeared that the Spanish authorities had abandoned Santa Rita del Cobre to its fate. Would the Apaches, now free to live and roam at will through these lands, allow the miners to remain in their midst? Only time would tell.

Following the signing of the 1810 treaty, the frontier was more peaceful for the next two years. Apaches who remained at peace received subsistence until lack of money again forced the government to cut off their rations. The Indians most troublesome to Santa Rita at this time were the Apaches living in the Sierra de Mogollón area. They knew these rugged mountains well, every rocky gorge, mountain pass, box canyon and water hole. After raiding villages and haciendas, the Indians retreated to their mountain stronghold with their stolen goods and animals They had no trouble eluding pursuing Spanish troops. Salcedo warned Baca that these bands were an immediate threat to the isolated mining camp. He advised the administrator to take whatever measures necessary for the safety of the residents. Baca formed a civil militia of miners who possessed some form of weapons. The militiamen prepared to defend the camp and furnish escorts for the mule trains. Baca posted guards, and permitted no one to go far from the mining camp.[11]

It wasn't long before the Apaches struck Santa Rita. Toward the end of

April 1812, José Baca reported, with a touch of irony, "the peaceful Apaches of this vicinity have left us without oxen, horses or mules." He went on to relate the details of this dilemma. During an April hailstorm, livestock belonging to one of the smaller mines got loose. When guards went to look for the strays, they discovered tracks that indicated six Apaches, camped near the property the preceding day, had removed the animals. They stole horses and mules from the pasture near Pachitejú on that same day. Baca formed a posse and followed the tracks toward the Mimbres River. They found no Indians or livestock; all had disappeared into the rugged mountains.

Four days later, Apaches ran off the only remaining horses. This time the tracks led toward the Ciénega de San Vicente. According to Baca, the culprits were Mogollón Apaches who had agreed to live in peace in their assigned territory. They were either shooting at the mining camp or leaving the area to raid in Sonora. He begged General Bernardo Bonovía, who had recently replaced Salcedo as Commandant General, to send additional troops from San Elizario and El Paso to put a stop to the depredations.[12] It was more than a year before effective campaigns by the military forced the Mogollón Apaches to again seek peace.

Mogollón chiefs El Fuerte (the strong one) and Mano Mocha (maimed hand) requested permission to establish rancherías along the Gila and Mimbres Rivers.[13] At first, José Baca was not in favor of the Apaches living in close proximity to the mining camp. These same Indians had raided Santa Rita livestock several times in past years. General Bonovía explained to Baca that it would be more advantageous to the government for the Apaches to live in Apachería rather than near the presidios and towns. Baca agreed that the Apaches would be better off establishing rancherías in their own lands. If they raised crops in the valleys they could store up provisions for the lean times. These Indians knew the Mimbres and Mogollón country well and would hunt and gather seeds and berries in the proper season. If they came in peace, the chiefs and their followers would be well received at the mining camp. He asked Bonovía to send an interpreter to help him talk with the new residents.[14]

Baca apparently became an agent at the peace establishment outpost. He distributed corn, cigarettes, and meat when they were available. The mining establishment probably provided these rations to promote friendly relations with a former enemy. The copper mines had no major problems with the Apaches for the next five years. However, Mogollón Apaches continued to raid in Sonora with great success. As their peace agreement was only with Chihuahua, they considered Sonora fair game. For the next two decades, Fuerte and Mano Mocha were in and out of the Santa Rita area. Their raids on Sonora and Chihahua continued, but they kept the peace with Santa Rita while Baca was there. At times they were helpful to the mining establishment. When the Coyotero Apaches of Sonora stole a drove of mules from Santa Rita, Fuerte is said to have led a raid on the Coyoteros in retaliation. Now and then they returned stray animals and passed on bits of information about activities of other Indians. Baca's ability to deal with neighboring Apaches allowed the Santa Rita mines to operate with little interruption.[15]

El Fuerte and Mano Mocha played important roles in the Santa Rita story for several years. Mano Mocha maintained peace with Santa Rita and continued to farm in the Mimbres valley until 1834. After this time, he and his band were considered hostiles. El Fuerte preferred Santa Lucia Springs (now Mangas Springs) in the Gila valley. There is much evidence to support the theory that El Fuerte became known throughout the southwest as Mangas Coloradas around 1842. If this is true, he continued to frequent the Santa Rita area after New Mexico became United States Territory. He was killed by American soldiers at Fort McLane (Pachitejú), only a few miles from the copper mines, in 1863.[16]

Mexico won the War for Independence against Spain and became a free nation in 1821. The following year, José Baca transferred to Chihuahua City where he continued to serve the Elguea interests. He remained with the firm for the next fifteen years as a valuable and trusted employee. Baca was a proficient mine administrator and maintained law and order in the mining camp. Because of his skill in dealing with the Apaches, Santa Rita was one of only a few mines that remained active through a decade of rebellion. When copper was in demand for coinage, Santa Rita del Cobre became the most important source of that metal. Although production records from 1810 to 1821 are incomplete, the Santa Rita mines produced over five hundred thousand pounds of copper in this period.[17]

7

Pablo Guerra and the Independence Movement

Pablo Guerra married María, the widow of Francisco Elguea on July 27, 1813, and assumed control of the prosperous commercial enterprise owned by the Elguea heirs. Mexico's long and bitter struggle for independence was now underway, and the country was in chaos. Santa Rita, like all of New Mexico, was far from the scene of combat, and largely unaffected by the fight for freedom. However, the rebellion did affect the economic health of Santa Rita del Cobre.

Pablo Guerra was a native of Spain, born in Zuaza, Castile, in 1786. He came to Chihuahua City in 1809, and worked as agent and cashier for the Elguea estate. After his marriage to María Elguea, he assumed management of the Chihuahua mercantile business, the Hacienda del Torreón of thirty-seven thousand acres, and the prosperous copper mines of Santa Rita del Cobre. Guerra soon became one of Chihuahua's leading citizens. He served on the city counsel and the board of public lands, acted as assistant magistrate for the Supreme Tribunal, and was a member of the Deputación de Minería.[1]

Francisco Jáurequi, manager and coexecutor for the Elguea heirs, was relieved to pass the responsibility on to Guerra. He found it difficult to honor all of Elguea's contracts and maintain his wide variety of business interests during these changing times. Elguea's contract to supply the presidio at Santa Fe, New Mexico was an example. Jáurequi said that the goods to fulfill this contract were very hard to obtain, especially during the insurrection. Mule trains transported the supplies to Santa Fe at great risk. In closing out his stewardship of the estate, Francisco Jáurequi made one final shipment from Santa Rita del Cobre, delivering 168,000 pounds of copper ingots to Mexico City.[2]

The Independence Movement and a decade of fighting created innumerable problems throughout Mexico. The unstable condition of the country prompted many wealthy Spaniards to return to Spain, taking their money with them. In 1814 alone, Spaniards took some twelve million pesos out of the country, and the exodus of Spanish families continued throughout the war years. Great amounts of money leaving the country as coins and bullion created a serious economic crisis for the government.[3]

Much of the fighting in Mexico's decade-long struggle for independence involved the rich agricultural and mining districts of central Mexico. Both royalists and rebels attacked and burned haciendas and drove off inhabitants. They killed cattle and sheep, destroyed fields and seized crops to prevent them from benefitting the enemy. Scarcity of agricultural products caused prices to rise sharply. Citizens living in the path of the rebellion experienced the same devastation that the northern frontier had lived with for years at the hands of hostile Indians.[4]

The keystone of New Spain's economy was the production of Mexico's silver and gold mines in the central part of the country. The revolution was disastrous for most of Mexico's mining districts. Many valuable mines, smelters and refining works were destroyed by insurgent activities and warfare. They wrecked machinery and pulled out shoring, which caused shafts and tunnels to collapse and mines to flood. Mines outside the battle zone suffered from lack of tools, supplies, food and workers. Transportation was disrupted when the government commandeered mule trains to carry war supplies. Mines were closed or abandoned, leaving workers without jobs. Some unemployed men joined the rebel forces. Others formed gangs of *bandidos* to roam the countryside, robbing and destroying property and terrifying the citizens. Silver mines still in operation found it impossible to get silver ingots safely to the mint in Mexico City. Even a strong military escort did not always prevent the bullion from falling into the hands of the insurgents. The mint could not produce coins without silver, and money was in short supply.[5]

In an attempt to make the conditions in the country better, the Viceroy declared copper money as legal tender for a low denomination trade media. People at the bottom of the economic pyramid needed coins of small denominations to avoid loss in exchanging silver coins of higher value. He ordered the mint to produce large quantities of copper coins of two, four and eight *marvedí*. (Two marvedí had a value of less than a United States copper penny.) Copper money was opposed by mercantile and financial interests. The

Viceroy assured circulation by ordering all royal officials, employees, and the military to receive one-third of their salary in coppers. Merchants reluctantly agreed to accept marvedí in partial payment for goods.[6]

The order to produce an abundant supply of copper coins presented a new problem for the distressed government. Before the insurrection, Mexico's principal source of copper was the Michoacán district of south-central Mexico. This was the home base of General Morelos, a leader of the independence movement. The Michoacán area was the scene of violence and destruction as royalist troops met rebel forces. Damage to mining properties was so extensive that the owners could no longer produce copper. Small copper mines in other parts of Mexico played out or closed for lack of tools, supplies, workers, and capital.[7]

This was a boon for the Santa Rita del Cobre mines. The Commandant General reported that these mines, far removed from the war, were still in good condition. They were well supplied with food and other necessities, and producing a considerable amount of high-grade copper.[8] There are indications that Pablo Guerra had advanced notice of the impending need for copper. José Baca, superintendent of the Santa Rita mine, already had a large amount of copper ore refined into bars and waiting. One week before the viceroy ordered copper coinage, the first of four long mule trains loaded with copper ingots was already on its way to the royal mint. Within three weeks more than 80,000 pounds of copper were in transit from Santa Rita to Mexico City.[9]

Pablo Guerra and José Baca continued to maintain a thriving business during the revolution. Between his political connections and his mercantile business, Guerra obtained the supplies and equipment necessary to keep the mine in operation. Santa Rita produced an essential commodity for the government. This eased the way to requisitioning scarce items for the mines, such as iron and steel for making and repairing tools. So great was the government need for copper that they relinquished all internal duties and taxes on the Santa Rita operation.[10]

Between three and four hundred persons now lived at Santa Rita. Guerra sometimes found it difficult to provide enough food for his workers and their families. Although produce was still available in the agriculture areas of the frontier, prices were high. Through the family mercantile business, however, he was able to meet their basic needs. The company store sold clothes, shoes, cigarettes and foodstuffs, but it was only open when supplies came in; there were no luxuries. Baca's good relations with Fuerte, Mano Mocha, and other Mogollón bands living in the Santa Rita district was an enormous asset to the mining camp. If the Apaches received occasional rations, they observed the peace with the mining camp. They even allowed the Santa Rita farmers to plant corn, wheat, frijol and vegetables in the Mimbres valley.

The copper mines continued to produce well, and Pablo Guerra's closely guarded mule trains were frequently on the road. Copper was of little interest to either Apaches or rebel forces. Therefore, transporting it was safer than was silver and gold. Guerra sold most of the copper from Santa Rita to the mint in Mexico City in 1814-1816 and 1821. Copper coins in the amount of 342,893 pesos were minted in those years.[11] He had no trouble finding other markets for the copper in the years the mint did not coin marvedí. In 1818 alone, Guerra sent more than 3,000 pounds of copper to Zacatecas; 700 pounds of smelted copper went to Durango, and 2,025 pounds to San Luis Potosí. He supplied copper to coppersmiths and craftspeople in Queretaro, Irapuato and Arispe, Sonora. The Santa Rita mines reported a net profit of fourteen to sixteen thousand pesos annually. When the high-grade ore became scarce for several months in 1819, however, Guerra invested most of the profits back into exploration and development. The following year Santa Rita was again in full production, sending more than eleven thousand pounds of copper ingots to Mexico City.[12]

Mexico became an independent nation with the signing of the Treaty of Córdova in August of 1821. Spaniards and their money continued to leave the country; others chose to remain in Mexico and participate in the formation of a new nation. Pablo Guerra was one of many Spaniards who took the oath of allegiance to the Sovereign Mexican Nation. By doing so, he could retain his political offices and protect his business interests for a time.[13]

Juan Luis de Hernández replaced Jose Baca as superintendent of the copper mines in 1822. Within a year or two, production at the Santa Rita mines began to decline. This was probably not the fault of Juan Hernández, but the results of political chaos and general disorder throughout the new nation. Commerce and industry languished. Although copper coins continued to circulate, they were never popular with the mercantile and financial interests. While copper was still used to give strength to silver pesos, it was no longer needed in large amounts at the mint. The copper market declined.

The central government had too many problems to deal with and not enough resources. The frontier received little attention or financial assistance. The states bordering Apachería now had the responsibility of maintaining the presidios and providing rations to the Apaches still living in the nearby peace establishments. When empty state coffers forced the governors to reduce rations to the Indians, the Apaches took up their old habits of raiding the settlements for their food supply. The presidios were undermanned, with poorly trained troops, shoddy weapons and an insufficient number of horses to mount an effective campaign against the Indians. When Navajos from the San Mateo Mountains raided Santa Rita in 1823, the Janos commander could do nothing to protect the camp. Eighteen men from his already meager force were in Sonora helping that commander in an offensive against Apaches. The remaining soldiers either were ill or had no provisions, uniforms, or horses.[14] Without the assistance of the military, the isolated mining camp was extremely vulnerable to Indian depredations.

The following year saw a general uprising from the Apaches when they did not receive meat with their ration of corn. The military commander put together a division of troops from the frontier presidios and sent them to Santa Rita to wage a campaign against the Indians. By the end of the year, Juan José Compá and a few leaders, probably Fuerte and Mano Mocha, came to Santa Rita to make peace. The commander could not promise rations regularly. Nevertheless, the Indians would receive corn and meat when it was available. The leaders and their bands returned to their rancherías and peace reigned for a short time.[15]

The brief interval of peace ended in the spring of 1826. The Apaches attacked the Santa Rita mining camp in force. They drove off all the horses and killed three men who were guarding them. When workers went into the forest to make charcoal for the smelting furnace, the Apaches ambushed them. This brought all mining operations at Santa Rita to a halt.[16] Pablo Guerra now faced a critical situation. Without strong military support he could not continue the business of mining and refining copper.

Independence was not the panacea for all of Mexico's ills. The nation was bankrupt, and political factions could agree on little. Spain refused to recognize the new nation and Mexico expected an invasion at anytime. At this time, Mexico was not a good place for wealthy Spaniards. There was growing hatred of the *gachupines* (Spaniards who emigrated to North America) who continued to hold important positions and profit from the government. Mexicans resented the Spanish merchants, who retained strong influence by lending money to the financially disabled government. The general public blamed the Spaniards for all their woes, and demanded that the legislature pass a law expelling them from the country.[17]

Pablo Guerra faced a dilemma. If the legislature enacted the Law of Expulsion, he could claim exemption and remain in Mexico. His wife and son were born there; he also operated a business that employed Mexicans. Nevertheless, did he want to stay in this country of political and economic turmoil? Guerra decided that he would take his family to Spain and establish a new home at Bilbao, on the Bay of Biscay near the French border. He began to arrange his business affairs to provide for the continuing prosperity of the family. Young Francisco Elguea would run the mercantile business founded by his father; his brother Mariano would continue to manage the Hacienda de Torreón.[18] José Baca was there to guide them.

Guerra needed to find someone to take over the Santa Rita del Cobre mines, either by lease or by purchase. This valuable property still produced a respectable amount of copper, in spite of a depressed market and frequent depredations by the Apaches. According to an American traveler in Mexico, Guerra realized a profit of one hundred talegas ($100,000) from the mines in the fourteen years that he was in charge of the Elguea estate.[19] Nevertheless, he would abandon Santa Rita if no one showed an interest in taking over the property. Guerra began to prepare the necessary documents required by law when a mine is abandoned.[20] Before he completed the legal proceedings, however, a party of American fur trappers, led by Sylvester Pattie, arrived at Santa Rita del Cobre and changed the course of events.

8

Sylvester Pattie & Son

The timely arrival of Sylvester Pattie and his son James, along with their trapping party, did more than prolong the life of Santa Rita del Cobre. James gave us a firsthand account of life at the copper mines in *The Personal Narrative of James O. Pattie,* published in 1831.[1] Readers often question the veracity of Pattie's narrative. His dates are a year off and his geography is at times confusing. He undoubtedly laced his story with fabrication for a more exciting and romantic tale. However, there is no doubt that James and Sylvester Pattie were at the copper mines. When combined with documented facts, Pattie's story becomes a very credible account of Santa Rita del Cobre in 1826 and 1827.[2]

Sylvester Pattie was born in 1772 in Bracken County, Kentucky. He grew up to become a farmer, miller, and skilled carpenter. In 1812 he sold his farm and moved his wife and children to Missouri. There he established a grist mill and a saw mill on the Big Piney River, rafting the lumber down the waterways to St. Louis. In 1822, Pattie's wife died of consumption (tuberculosis), leaving Sylvester with eight children to raise. Sylvester lost interest in the lumber business after his wife died. He was restless and felt the need to try his hand at some new venture.[3]

By this time, reports of successful trading and trapping expeditions to New Mexico had reached Missouri. The fledgling Mexican nation was dealing with an economic crisis brought on by a decade of war. In an effort to bolster Mexico's sagging economy, the government did away with the old Spanish policy banning foreigners from their country. Up until this time, any foreigners crossing the border into New Spain were summarily arrested by Spanish soldiers. Now the Mexican borders and ports were open to traders from the outside world. Aliens received a warm welcome from New Mexicans eager to trade. When word of this development reached the United States, merchants organized mule trains loaded with trade goods and headed for Santa Fe.

Each year more caravans followed the route soon known as the Santa Fe Trail. When the New Mexico markets became saturated, the foreign merchants traveled the Camino Real south into Chihuahua, Durango and Sonora. With consumer goods no longer available from Spain, frontier merchants were eager to purchase American products which they could then retail at a profit. In exchange for all kinds of manufactured goods, the Americans received silver bullion, mules and fine-quality beaver skins which were much in demand in the United States and Europe. Beaver pelts brought $5.00 a pound in St. Louis. Trappers, eager to test the rivers of the west for this fine beaver, soon joined the trade caravans.[4]

Sylvester Pattie decided to seek his fortune in the new country of Mexico. He sold his lumber business and invested in trade goods, trapping gear, and supplies. At Council Bluffs, Sylvester and his son James joined a large party of trappers sent out by Bernard Pratte & Company of St. Louis. In late October 1825, the party reached New Mexico, where they had trouble obtaining a trapping license. The governor of New Mexico, concerned that foreigners would soon deplete the beaver population, issued a trapping license only to Mexican citizens. One way to get around this restriction was to trap under a license issued to a Mexican citizen and take the man with them to learn the art of trapping. Sometime the trappers obtained a license to trade with the Indians, trapped illegally, and passed their furs off as the results of a successful trading venture.[5]

Pratte arranged with Governor Antonio Narbona for a license. He divided his company into three groups, and sent them out in different directions. One party of seven men, led by Sylvester Pattie, headed for the Gila River. They journeyed down the Rio Grande valley to the vicinity of the present town of Truth or Consequences. From here the party turned west to pick up a rough trail through the mountains known today as the Black Range. After four days, they reached Santa Rita del Cobre. The hunters spent one night at the copper mines, where they hired two men to serve as guides before setting out in a northwesterly direction for the Gila. They found plenty of fine black beaver on the Gila River and its tributaries. All went well until the Indians stole their horses. The trappers were forced to cache their furs, planning to retrieve them later.

After a disastrous four months in the wilderness, the exhausted and half-starved men returned to the copper mines. The miners received the bedraggled party with much enthusiasm and probably with some surprise, as they did not expect to see the trappers again. Apaches recently made their annual spring raid on the Santa Rita livestock with great success, stealing most of the horses and mules. They ambushed the charcoal makers every time they went into the forest. Smelters were idle, and all mining operations were at a standstill.[6]

Superintendent Juan Luis Hernández (Pattie called him Juan Onís) willingly supplied the immediate needs of the trappers. Pattie thought him "a gentleman of the highest order." Juan Luis accompanied James and four others to Santa Fe to pick up supplies and purchase horses. Hernández also furnished provisions for the trappers, and armed ten of his laborers to accompany the small party to retrieve their cache of beaver skins. This hazardous journey was all for nothing. When the trappers arrived at the place where they left their furs, they discovered that the Indians had stolen the lot.[7]

The discouraged trappers returned to the copper mines. Hernández offered to pay them a dollar a day if they would remain at Santa Rita until the end of the year guarding the charcoal makers while they worked in the forest. Sylvester and his party decided to accept the offer. They refused the salary, however, because the superintendent had already done so much to help them. This was a pleasant interlude for the Americans. They hunted the plentiful deer, bear and turkey, and explored the country when not on guard duty.[8]

Sylvester Pattie established good relations with the Apaches. He met with chiefs Mano Mocha (James calls him Mocho Mano), Fuerte, and others who lived near the mines. The Apaches declared that although they had long been at war with Spaniards and Mexicans, they had no bad feelings against Americans.[9] Sylvester convinced the Apaches that Americans were strong, brave, and excellent marksmen. They could raise a sufficient force to make war if the Indians did not allow them to work the mines without fear of attack. According to James, this either impressed or intimidated the

Apaches, for they agreed to allow the miners to resume work. Mano Mocha offered Sylvester a large tract of farmland in the Mimbres Valley. Spaniards tried to farm in this valley several times in years past, but the Apaches always killed the farmers and destroyed the crops. Now the Indians promised they would disturb nothing belonging to the mines. They even agreed to return strayed or stolen stock.[10]

The Apaches were very friendly with the Americans, coming to the mining camp often with deer and turkey to sell. Juan Hernández always traded with them to maintain good relations, even when he did not need the game. He knew the presence of Sylvester and his sharpshooters was all that kept the Apaches at peace. Pablo Guerra was pleased to have Santa Rita back in business. He was on the brink of abandoning the property. Now the mining and smelting operation resumed on a grand scale. Between January and September of 1826, Guerra's mule trains carried seventy-three tons of copper ingots and plates to markets in Mexico City.[11]

The copper mines became a familiar landmark on the route to the Gila as more trappers headed west. In the fall of 1826, James Baird reported that more than one hundred foreigners trapped beaver on the Gila and its tributaries.[12] Baird, a fellow American, came to the southwest in 1812. He became a naturalized citizen of Mexico and a resident of El Paso del Norte. Baird, a trapper himself, expressed concern about too much competition from the "foreigners." He declared that over the past eighteen months small illegal groups had taken more than one hundred thousand dollars worth of prime furs from New Mexico.[13]

As many trapping parties passed by the copper mines on their way to the beaver grounds, the Patties prepared to join the fall hunt. If the Americans left Santa Rita, peaceful relations with the Apaches were sure to end. Pablo Guerra proposed leasing the mines for five years to Sylvester and such partners as he selected. As James described the terms, they were more than generous. The lease fee would be $1,000 per year. Guerra agreed to furnish provisions for the first year, at no cost to the Americans, and to pay for any improvements they made on the property. Sylvester accepted this offer. The lease probably began the first of October, as Guerra made his last copper shipment of record in September.[14]

James Pattie said that his father had partners who worked the copper mine with him, but he failed to name them. Perhaps Nathaniel Pryor and James Kirker had an interest in this venture. Both men came west, in the fall of 1825, to trap beaver and trade with the Indians. They were with the Patties when they left the mines in 1827. Pryor accompanied Sylvester and James to California. Kirker, who became a famous Indian fighter, returned to Santa Rita. He made the copper mines his headquarters for the next ten years.[15] Sylvester and his associates had no mining experience; their duty was to maintain peace with the Apaches and guard the mining operation. Sylvester hired an overseer to run the mining operation and handle the clerical work. The new superintendent spoke several languages, and his ability and apparent good character impressed Pattie.

James showed a decided lack of interest in the Santa Rita mines by this meager description: "Within the circumference of three miles, there is a mine of copper, one of gold, and a claim of silver which is not worked as not being so profitable as either the copper or gold mines; there is also a cliff of lodestone." Young James preferred the adventurous life, and soon he was off on another trapping expedition.[16]

The Mexican archives fail to mention Americans in charge of the Santa Rita mines during this period. Other sources verify their presence, however. An Englishman, R.W. H. Hardy, toured Mexico in 1826 -1828. When he passed through Janos, he heard of the Santa Rita del Cobre mines and later wrote this description: "This latter is a Real of copper. There is also a *creadero de oro* (place where gold originates) near it; but the Indians, who formerly gave it to a Spaniard, since deceased, would not suffer it to be worked till within the last year. An American has succeeded in conciliating this tribe; and it is said he obtains considerable emolument from the metal."[17]

Sylvester Pattie and his associates did prosper at Santa Rita del Cobre. Besides mines of copper and gold, they conducted a thriving mercantile business at the company store. Often, the workers preferred to take part of their wages in the form of merchandise. Pattie also purchased wine and whisky from El Paso, and sold it at the mine for $1.50 a pint. James said that his father made at least 200% profit on the mercantile end of the business. Sylvester established a ranch for his livestock in the Mimbres valley, and planted crops on the land Mano Mocha gave him. A good yield enabled him to put aside a supply for the future.[18]

Toward the end of the first lease year, Sylvester sent his trusted clerk to Santa Fe with thirty thousand dollars in gold bullion to purchase merchandise for the store. After a month with no word from the clerk, Sylvester dispatched James to investigate. The clerk had not been seen. He either absconded with the money, or someone robbed and killed him. This was a serious setback for Pattie & Company, with a whole year of work wiped out. Sylvester sent his son to Chihuahua to seek help from Pablo Guerra.[19]

When James Pattie arrived in Chihuahua, he found the Spaniard preparing to leave Mexico. Guerra hoped that Sylvester Pattie would continue to lease the Santa Rita mines, or even purchase the property. He was disappointed to find that he still had a copper mine on his hands. Concerned with the uncertainty of his own future and the welfare of his family, Guerra could do nothing to aid the Patties.

The only solution open to Sylvester and his associates was another trapping expedition. If they could make a good haul of beaver, they might recoup their losses. Therefore, Sylvester closed out the business at Santa Rita, dividing the little profit left between himself and his associates. He outfitted his party with traps and supplies for the coming expedition. On September 22, 1827, he obtained a permit from the governor at Santa Fe, allowing him to go to Chihuahua and Sonora to trade.[20]

Sylvester, James, and others from Santa Rita joined a trapping party headed for the Colorado River. After the usual troubles and hardships which seemed to follow the Patties, they reached California. They were suspected of being spies and thrown in prison, where Sylvester died May 24, 1828. James finally made his way back to the United States in August of 1830. All he had to show for almost five years of wandering in the west were his memories of a grand adventure. With the help of Cincinnati editor Timothy Flint, James turned his odyssey into a book that is now a Western classic.

9

A Takeover by McKnight and Courcier

Robert McKnight and Stephen Courcier took over the Santa Rita property in the early fall of 1828. Pablo Guerra had lately departed for Spain, leaving the mines in the hands of his agent, Luis Durasta. McKnight and Courcier were eager to acquire this valuable property. They mined Santa Rita copper for the next ten years. These were years filled with big profits hampered by many difficulties.

Robert McKnight came to New Mexico in 1812, from St. Louis, Missouri where his family operated a mercantile business. At that time, Spain permitted no foreigners to cross their colonial borders. When word of the 1810 revolution in Mexico reached St. Louis, the Americans concluded that Mexico was now open to foreign trade. McKnight and eight others organized an expedition to New Mexico to take advantage of the new market. Pack mules were loaded with trade goods and provisions valued at $10,000.[1] The traders learned of their mistake when they arrived in New Mexico. Mexican authorities arrested the party as spies and confiscated their trade goods. Soldiers escorted the Americans to Chihuahua, where they spent two years in prison. The Mexican government released the prisoners under bondage in 1814, and the Americans went to various places in Chihuahua and Durango to work for the men who provided their bond money. McKnight's benefactor was a merchant from Guarisamy, some two hundred miles from Durango near the border between Nueva Vizcaya and Sinaloa. This area, in the Sierra Madre Occidental, was one of the most productive silver mining districts in Mexico at that time. McKnight did well in Guarisamy. He worked in the general store and is said to have received part of the profits. Robert acquired a wife while there, and a general knowledge of the mining business which he put to good use at Santa Rita del Cobre in the years to come.[2]

When the King of Spain ordered the release of all foreign prisoners in 1820, McKnight chose to remain in Mexico for two more years. When he returned to Missouri, in 1822, he sought compensation from the United States government for the trade goods lost to the Spaniards in 1812. The government turned down his petition. Trying to improve his finances, he and his brother John joined a disastrous trading expedition to Comanche country, in which John lost his life. Nothing seemed to go right. Robert decided to return to Mexico. He said, "there is a better chance of obtaining justice from the Mexicans, scoundrels as they are, than from my own government. I will go and recover as a citizen of Mexico what I lost as a citizen of the United States." On May 16, 1825, he joined a caravan of merchants bound for Santa Fe. There is no indication that he ever visited the United States again.[3]

Information about McKnight's partner, Stephen Courcier, is scarce and contradictory. He was of French ancestry, born around 1776 in Philadelphia. R.W.H. Hardy, a contemporary author touring Mexico, met Courcier in Chihuahua in 1826. Hardy said the French gentleman came from New Orleans by way of Texas. Another source claimed that he came to Mexico from St. Louis with Robert McKnight in the 1825 trading caravan. Courcier had money to invest. He applied for Mexican citizenship, opened a store and often served as interpreter and translator for the government. Courcier became an important figure in politics, merchandising, and mining on the frontier.[4]

When Robert McKnight and Stephen Courcier learned that the owner of the copper mines had left the country, they headed for Santa Rita, planning to take over this lucrative property as an abandoned mine.[5] In Janos, they met Guerra's agent, Luis Durasta, who informed them that Santa Rita was not an abandoned mine. He threatened them with immediate litigation if they persisted in their attempted takeover. McKnight and Courcier admitted having doubts about the strict legality of their intended denouncement. They were willing to sign a five-year lease, paying one *quintal* (one hundred pounds) of copper per week for the right to work the Santa Rita mines.[6]

The lease contract for Santa Rita del Cobre was probably similar to a lease on the same mines issued to Francisco Siqueiros thirty years later. The lease stipulated that any improvements made on the property reverted to the owner at the end of the lease period. A lessee had the right to cancel the contract if copper production failed to meet expectations, or if hostile Apaches forced the tenant to give up the mine. Stephen Courcier held the lease and conducted the business affairs of the partnership from Chihuahua City, while Robert McKnight took charge of the mining and smelting operation. This arrangement produced a successful and long-lasting partnership.[7]

The Santa Rita copper mines were in poor condition when McKnight and Courcier took them over in 1828. There is no evidence of any mining carried on at Santa Rita in the year following Sylvester Pattie's departure. McKnight and Courcier found it necessary to invest a large amount of capital in the property in order to bring it back to full production. They planned to use a large work force and the best mining methods known at the time. McKnight brought his nephew John from Missouri to work with Courcier in the mercantile and supply business.[8]

Courcier ordered mining supplies and equipment, and several heavy freight wagons from St. Louis to use for hauling the copper ore to market. The trail to Janos, however, required extensive work to make it a passable road for the wagons. There were several advantages to using wagons rather than mule trains. A team of ten to twelve mules pulling a large wagon could haul more freight than the same number in a pack train. Mules were also better than oxen at pulling the wagons through sandy washes. A wagon, once loaded, was good for the entire trip, while mules must be unloaded at each camp and reloaded the next morning before getting underway. This was a time-consuming business. Wagons could make twenty miles on a good day, and the merchandise was more protected in the wagon trains that were always escorted by well-armed guards.[9]

A colony of a hundred workers and their families came to make their home at Santa Rita del Cobre; in time the population increased until it numbered around four hundred. McKnight and Courcier tried to provide the employees and their families with what they needed to make life tolerable in this isolated camp. Workers were encouraged to build small adobe homes and raise vegetables in garden plots along the creek. They planted an orchard of peach and apple trees and put in a vineyard. Farmers raised wheat and corn in the Mimbres valley when the Apaches were peaceful. All other supplies came from Chihuahua, Sonora, or the United States. McKnight gave his workers a basic diet of corn, frijoles, and salted or jerked beef, with fresh beef or mutton when available. The company store offered a variety of goods. Besides foodstuffs and staples, a typical stock included wool and flannel blankets, garters and sunbonnets for the ladies, with shirts, pants, and jackets for the men. The store also carried stockings and shoes, shaving knives and soap. For the seamstress and tailor there were bolts of cotton and woolen cloth, sewing needles, thread and buttons, ribbons and braid. Workers could charge their purchases until the next payday. McKnight arranged for Father Rafael Echeverría, chaplain of Janos presidio, to visit the camp every six months and minister to the people living there.[10]

Several Americans found employment at the copper mines. Lewis Dutton,[11] Henry Corlew,[12] James Buchanan[13] and James Kirker[14] came to New Mexico around 1825. These young men had several years of experience as trappers and traders by the time they joined McKnight at Santa Rita del Cobre. At first Dutton, Buchanan and Corlew worked for themselves extracting copper under a sublease from several shafts on the property. Dutton later became the foreman of the mines and Buchanan and Corlew were hired on as wagonmasters.

James Kirker was a family friend who once worked for Robert McKnight's brother in St. Louis. Kirker was at Santa Rita in 1826, and he left there with the Patties in the fall of the following year. It is entirely possible that Kirker informed McKnight and Courcier that Pattie had abandoned Santa Rita, and encouraged them to take over the valuable copper mines. Kirker was in and out of Santa Rita during the next ten years. He continued to trap in the winter and protect the copper mining operation in the off-season. These four Americans remained at Santa Rita the entire time that McKnight operated the mine. Kit Carson worked as a teamster at Santa Rita for a short season, but did not like the job and soon returned to Taos. He became famous as a trapper, Indian fighter and guide. His work often brought him along the trail through Santa Rita.[15]

Several other American trappers made their headquarters at the mining camp over the next ten years. Harvesting beavers was a profitable business. The pelts sold for $4.00 to $4.25 per pound in Santa Fe and $5.00 to $6.00 in St. Louis. The Mexican government still refused to issue a trapping license to foreigners. Nevertheless, American and French trappers continued to harvest the high-quality beaver skins from Mexican rivers and streams. However, if the trappers appeared in Santa Fe with a large amount of furs, and no trapping license, they faced arrest and confiscation of their pelts. The trappers soon found a way to get around this restriction by ending their beaver hunting season at Santa Rita. They hid their furs and trapping gear in deep, abandoned mine shafts on the property. The trappers then went to Santa Fe where they had no trouble getting a license to trade. Returning to the copper mines, they retrieved their furs and took them to Santa Fe in small bundles, declaring them as the results of trading with the Indians. Santa Rita was not under the jurisdiction of Santa Fe, therefore the authorities paid little attention to covert activities that occurred in this out-of-the-way place.

These hardy frontiersmen and their rifles were important to the success of the Santa Rita mines. McKnight willingly hired the men in the off season to protect the miners, guard the livestock, and escort the wagons along the hazardous road to Chihuahua.[16] Santa Rita del Cobre never enjoyed the popularity of Taos as a gathering place for trappers; nevertheless, it offered a safe haven between beaver seasons for those operating outside the law.

The key to the success of the Santa Rita mining operation was good relations with the Apaches living in the area. Jose Baca achieved this in 1814 when he allowed Fuerte, Mano Mocha and their bands to establish rancherías in the Santa Rita district if they remained at peace. Baca administered a form of the peace establishments maintained at the presidios, handing out occasional rations and gifts, and encouraging the Apaches to raise crops in the Mimbres and Gila valleys. After Baca left at the end of 1821, the local bands became more troublesome. Sylvester Pattie also established good rapport with the Indians. Although the American trappers were well armed and skillful with their rifles, they offered friendship as long as the Apaches kept the peace. Robert McKnight used the same methods with some success. The Apaches trusted him to treat them fairly, and they wanted his presence whenever they were conducting peace negotiations with the Mexicans.

Maintaining friendly relations with the Apaches had its dark side, however. As beaver became scarce from over-harvesting, some enterprising Americans discovered that trading with the Indians for illegal merchandise was a profitable business. In exchange for good American rifles, fine-quality gunpowder, and potent liquor, the Americans got the spoils of Apache raiding—horses, mules, stolen goods and sometimes human captives. As early as 1812, the commander of a frontier presidio accused a Santa Rita resident of selling gunpowder to the Apaches. Santa Rita del Cobre provided the perfect setting to carry on this illegal exchange of goods because of its isolated location.[17]

This clandestine bartering with the enemy became a major problem soon after McKnight took over administration of the mining operation. By providing a market for stolen property, the traders encouraged the Apaches to continue raiding in Chihuahua and Sonora. Mexico's secretary of state demanded that the United States' foreign minister to Mexico do something to stop the Americans from carrying on this illicit trade. Robert McKnight and James Kirker were accused several times of participating in this lucrative sideline at Santa Rita, but nothing was ever proved against them.[18] Trade with the Apaches did not guarantee their friendship; they might raid one week and trade the next. But what better way to replace stolen livestock?

While Robert McKnight dealt with the Apaches and produced an abundance of copper ingots, Steven Courcier established markets for the copper. Chihuahua desperately needed money to bring economic stability to the state, strengthen the crumbling presidial system and finance military operations against the Apaches. Chihuahua Governor José Isidro Madero concluded that if the state (created in 1824) could mint its own copper money and control the value, it would help the economy. The governor ordered the Chihuahua mint, which closed during the War for Independence, restored and put back into service.[19] As Santa Rita was the only active copper mine in the state at this time, Courcier found a ready market supplying the government with all the copper McKnight could produce. By the following year, the state was turning out copper coins in the form of the cuartilla with a value of one-fourth real, the tlaco worth one-eighth real, and the pilon of one-sixteenth real.[20] Mexico City also offered a ready market for copper, although it was a long journey from Santa Rita.

The copper mines did extremely well under the management of McKnight and Courcier. By the time their five-year lease was up, Courcier controled the Chihuahua copper market. The success of the McKnight-Courcier operation drew much criticism from José Agustín de Escudero, a leading Chihuahua legislator and pamphleteer. Escudero said that Courcier's copper monopoly had caused the price of copper to more than double. It had "destroyed the mills in Chihuahua, ruined the coppersmiths and suspended the trade in copper utensils and other goods used in the homes and kitchens." Escudero was fair enough to add: "What Senor Courcier is doing in maintaining his business in the middle of hostile Indian country and with the other hardships demanded by working the mine, smelting and transporting the ore, could not be done by any of those who secretly view his profits with envy."[21]

Juan Álvarez, Pablo Guerra's man of business in Chihuahua, wanted to give the contract to someone who could make a profit and still sell the copper to the public at a more equitable price. The agent was unable to find such a person, however, nor could Escudero interest men of means in forming a mining company to take over the property. Alvarez was obliged to renew the lease with Courcier for two years, doubling the fee to two hundred pounds of copper a week. With the copper mines at the peak of their prosperity, Courcier did not hesitate to accept the terms of the new lease.

10

A Struggle for Survival

Ten years after Mexican independence, the country still struggled to establish a stable government while various factions vied for power over a bankrupt nation. The central government could give little guidance or financial assistance to the northern states and territories. This in turn forced the frontier states to deal independently with the problems of decaying military garrisons, agitated Apaches, and a shrinking economy.

The Apaches created many frontier problems for which there were no easy solutions. In Spanish Colonial times, the government had a strong, well-organized army. With intensive military campaigns, they persuaded many Apaches to live in peace near the presidios, where they received rations, gifts and various benefits. This very costly arrangement began to fail during the War for Independence. The peace establishments were now the responsibility of the frontier states. The Mexican government, impotent and deeply in debt, did not lift a hand to help the states deal with the situation.

Chihuahua and Sonora could no longer afford the expense of maintaining several hundred Apaches living at the peace establishments. The Indians once received corn, meat, sugar, salt, tobacco, blankets, gifts, and rewards for good behavior. The overwhelming expense of funding the peace establishments led to budget cuts and reduction of rations until the peaceful Apaches received only a small weekly measure of corn. In May of 1831, the government stopped even this meager ration. The angry Apaches left the peace establishments in open rebellion and joined their compatriots in raiding the Mexican settlements to avoid starvation. Incessant depredations again made the roads unsafe for travel, disrupted communications, destroyed crops, stripped ranches of their stock and terrorized the inhabitants of isolated villages, haciendas and mines.

The Mexican army was ill-equipped to deal with the Indians. Frontier presidios were undermanned, while vagabonds, petty criminals and peons found themselves conscripted to make up the ranks. Sometimes the state treasury had no money to pay the troops and provide rations for them and their families. Desertions became common, leaving the presidios in a vulnerable predicament. They could hardly defend their own horse herds from Apaches, much less mount an active campaign against the Indians. In June of 1829, Colonel Simón Elías González, Commandant General of Chihuahua and New Mexico, ordered the Janos commander to furnish a corporal and five-man escort to accompany the supply train to the Santa Rita mines. The Janos commander replied that not only was there no corporal to send, but the troops desperately needed horses and mules to carry out their regular duties. Again, the Apaches had stolen all the animals belonging to the presidio.[1] Santa Rita del Cobre, with over four hundred people to provide for, was left to depend on its own initiative and resources to survive.

Robert McKnight did everything possible for the safety and welfare of his Santa Rita people. He borrowed twelve English carbines from the Janos presidio and organized his miners into a civil militia to guard the camp. He posted lookouts to watch for Apaches, and the miners worked with their weapons nearby. Horses and mules pastured at Pachitejú ranch, several miles south of the mining camp, were prime targets for the Apache raiders. Three or four stockmen guarded the herd, offering a weak defense against large bands of marauding Apaches. Courcier's freight wagons brought in large quantities of food and supplies every three or four months, and carried the copper bars to market. McKnight provided ten to fifteen armed guards to escort the wagon train along the hazardous route. On the return trip, the wagons carried up to 15,000 pounds of copper ingots to the mints in Chihuahua and Mexico City.[2]

In 1831 Colonel José Joaquín Calvo came out of retirement to serve as Commandant General of the state of Chihuahua and the Territory of New Mexico. Colonel Calvo was a competent officer with many years of frontier experience. He had been Commandant General of Sonora and Sinaloa until his retirement in 1828, and he was well aware of the problems he faced. Calvo was determined to restore peace and prosperity to the northern states and territories.[3] Unable to pacify the Apaches without offering them rations and gifts, Calvo declared war on them. He placed the entire state on alert. Calvo ordered administrators of haciendas, ranches, and villages to arm their people with a weapon of some sort (a rifle, a lance, a bow and arrow, or even a slingshot with a pouch of rocks), or face a fine of twenty-five pesos.[4] The governor activated three companies of state militias. A call went out for horses and money in order to mount and provision the troops. The continued success of the copper mines depended on keeping the Apaches under control. Stephen Courcier made a generous donation of five hundred pesos, and twenty-eight men of the Chihuahua civil militia and ten soldiers from Janos went to protect Santa Rita del Cobre.[5]

Colonel Calvo organized a general campaign against the hostile Apaches in the spring of 1832. Soldiers from the presidios of Janos, Carrizal, and San Buenaventura, and all civilians whom they could mount and arm were placed under the command of Captain José Ignacio Ronquillo.[6] They would work with forces from Sonora and New Mexico. Such a mighty army was bound to bring the Apaches back to peace. After establishing headquarters at Santa Rita del Cobre, the troops made a reconnaissance of the Mimbres, Gila and Mogollón mountains in search of Apaches. Only one decisive battle occurred. On May 23, Lieutenant Ronquillo and 138 soldiers met some 300 Apaches led by Fuerte, Mano Mocha, and Pisago Cabezón, on the Gila River. The military forces killed twenty-two warriors, wounded fifty-one and took two prisoners. They recovered one hundred forty stolen horses. Sonoran troops won a decisive battle a few weeks later. Following their usual pattern after engaging in battle with the military forces, several Apache leaders came to Santa Rita del Cobre to discuss peace. Captain Ronquillo arrived with a large company of soldiers from Janos to represent Commandant General Calvo in the negotiations.

The twenty-nine chiefs who signed the treaty in August of 1832 promised to live in peace in the territories assigned to them. Juan José Compá was appointed leader of Apaches formerly living at peace establishments near the presidios. Their land was from the presidial line to the Burro Mountains to La Casita de Gila with intervening mountains up to Santa Lucia (Mangas Springs). Apaches who had rancherias in the Mimbres and Gila areas were granted land from the Mimbres Mountains westward to Negrito (near San Francisco River), including the Mogollon Mountains. Fuerte was their chosen leader, with headquarters at Santa Rita. The third group was established on the Sonora frontier under the leadership of Chief Aquien. The Apaches agreed to live in their assigned areas, maintaining an orderly way of life. They were not allowed to leave these areas without the permission of the Commandant General. They promised to return all horses they had stolen from the ranches, haciendas and presidios, and to make no more raids on Chihuahua and Sonora. The Apaches selected Juan José Compá as "First General" of the tribes on the frontier.[7]

Juan José Compá became a key figure in the Santa Rita story for the next few years. He was the son of El Compá, a Chiricahua Apache who worked closely with the Spaniards when the first peace establishments were set up in the late 1780's. At one time, the Compá family lived within the walls of Janos presidio. As a young lad, Juan José attended school with the military children where he learned to speak, read and write Spanish. When he grew older and formed his own band, Juan José maintained good relations with Mexican authorities and often served them as a scout, informant and interpreter.[8] The Commandant General expected Compá to use his influence with the recalcitrant Apaches to persuade them to abide by the terms of the latest agreement.

A brief period of peace and prosperity followed the signing of the treaty at Santa Rita in August of 1832, but it lasted less than six months. The government refused to restore the ration system, hoping to turn the Apaches to agriculture rather than stealing livestock for their subsistence. This led to great dissatisfaction among the Apaches, who expected this bounty as one of the rewards for agreeing to remain at peace. Incidents began to occur along the frontier such as the following:

All appeared peaceful at Santa Rita on a morning in May of 1833. Apache leaders Mano Mocha, Caballo Ligero, and Antonio Pluma and their bands prepared their lands in the Mimbres valley for cultivation. Mexican citizens Nicolas Ruiz, Juan Rodriquez and a man called Clemente also had farm land in the fertile valley and lived in harmony with the Apaches. Chirimi, an Apache whose ranchería was on the Mimbres, appeared at Santa Rita. He asked for a permit for himself and his family to go to Carrizal for the harvesting of crops. Chirimi always abided by the terms of the latest peace treaty and got along well with the people of Santa Rita. His sons worked in the mines and were always well treated. On this day in May, he sent two of his men to the Mimbres River to get some tools. It was later reported that they went to the river to warn Mano Mocha and the others that trouble was coming.[9]

Shortly after Chirimi's departure, the sister of Juan José Compá was found dead a half mile from the camp, on the road to the placers. Was this a vendetta against Juan José? Many Apaches resented his position as "General," and accused him of betraying their plans to the government. Juan José claimed that his life had been threatened. Later in the day, McKnight's horse herd at Pachetijú was raided. One guard was killed and some of the stock taken. McKnight suspected Chirimi's band of both crimes. He gathered an armed posse and went in pursuit. The trail led toward Santa Lucía where El Fuerte and others had their rancherías, but he found neither the culprits nor any of the stolen horses. Fearing more trouble, the Mexican farmers from the Mimbres valley abandoned their land and fled to Santa Rita. Apache farmers moved their families into the Mimbres Mountains. There they reportedly joined a large group of rebels from Carrizal, who gathered in the mountains with the intention of attacking the mining camp.[10]

Work at the mines came to a stop as McKnight organized his civil militia company to defend the camp. He sent a message to the Commandant General informing him of the imminent danger and asking for more help to guard the wagon trains bringing in supplies "upon whose existence more than four hundred people in this camp are depending."[11] Ten soldiers from Janos were assigned this duty. Cutting off the food supply offered the surest way to victory for the Apaches. This time, however, the expected attack did not come. Well aware of the preparations being made for the defense of the mining camp, the Indians turned their attention to the south and west, causing ruin and destruction throughout Chihuahua and Sonora. The miners went back to work, and the McKnight Courier operation continued to show a profit in spite of the havoc caused by Apache depredations.

The Commandant General sent a message to Juan José reminding him of his obligation to see that the Apaches complied with the terms of the peace treaty negotiated the previous year at Santa Rita.[12] Juan José was in a dilemma. He wanted to retain the important position of General to which his people had elected him. At the same time, he wanted to continue receiving the benefits derived from the government of Chihuahua as a trusted intermediary. He was afraid to go near the presidios lest the Apaches think he had broken faith with them. He could do little to prevent the Indians from leaving their assigned areas.[13]

Ten soldiers from the Janos Company remained on detached service at the copper mines. This small company was hardly enough to furnish a strong escort for the wagons and mule trains, much less patrol the area, protect the miners, and guard the animals. In the summer of 1834, the Apaches renewed their effort to dislodge the miners from Santa Rita del Cobre. More troops were sent from the presidios to reinforce the mining camp. The Apaches showed their scorn for the soldiers by running off the entire military horse herd of 160 animals pastured near the mines. Four soldiers deserted following this attack; they had been responsible for guarding the herd, and knew their punishment for failure would be severe.[14]

The Mogollón Apaches were also creating havoc in Sonora with their frequent raids. From their rancherías along the San Francisco River, they moved south and set up their base camp in the Chiricahuas to gather agave. From there, they raided Sonora settlements, sending the stolen horses, mules and cattle north to their rancherías. Sonoran officials accused the Chihuahua government of being too quick to offer the Apaches a peace treaty, knowing the Indians had no intention of abiding by the terms of the agreement. Furthermore, they said the Apaches were getting their lead and gunpowder from Santa Rita del Cobre in exchange for mules. At one time the Sonorans even considered taking legal action against Robert McKnight for arming the rebel Apaches.[15] This was not the first time the Americans at Santa Rita del Cobre were suspected of exchanging firearms and ammunition for mules. The Sonorans were probably justified in their accusations, but trading with the Apaches did not protect Santa Rita del Cobre from their raids.

The extent of the Apache preparations for raiding and warfare were revealed in a report from Jose Madrid, a young lad who escaped from the Apaches after spending several months in their camp as a captive. Madrid gave the authorities an alarming picture of more than five hundred hostile Apaches who had joined forces and were living in rancherías on the San Francisco River near its junction with the Gila (near present-day Clifton, Arizona).

The Apaches had acquired large herds of cattle, horses, and mules from their raids. The Indian's horse herd was far superior to that of the military, as many of the animals were stolen from the presidios. The Apaches were skilled and daring horsemen and warriors by nature. Madrid said that the horses were kept in the roughest places and well maintained. The warriors exercised their mounts daily to accustom them to the rigors of the plundering forays and the noise of firearms. Most of the Apaches carried rifles of good quality and had plenty of gunpowder obtained in trade with the Americans. On the other hand, the Mexicans made do with insufficient firearms in poor repair and a meager supply of ammunition.

Jose Madrid also reported that Juan José Compá was a frequent visitor to the Indian camp. Compá was invited to join the rebels, but he refused and urged the Indians to consider the peace terms being offered by the Mexican authorities. Some of the older Indians were in favor of peace, but the younger ones and the women were against it. Bands of one hundred or more Apaches went out to raid in Chihuahua and Sonora, while others went to New Mexico and brought sheep back to the camp. Madrid warned that the Indians planned to destroy the presidios at Carrizal and Janos, then they would step up their attacks on the rest of Chihuahua.[16] Madrid's warning came too late. On September 28, 1834, a force estimated at between two hundred and three hundred Apaches struck Janos, killing two men and running off more than one hundred thirty horses from the presidial herd.

Santa Rita del Cobre was also one of the Apache's prime targets. They appeared at the mining camp in force a few days after they struck Janos. The military records yield little information about this incident. There is no mention of either miners or Apaches killed or wounded. The Indians avoided battles if they could get what they wanted in other ways, and they wanted the Mexicans and Americans out of Apachería. They still appeared to trust McKnight, even in this tense situation. He and his people had always been good to the Apaches who lived in the Santa Rita district. The miners traded with the Indians and employed a few to work at the mines. Apache and Mexicans labored side-by-side on the Mimbres farms. It is entirely possible that the Apache leaders allowed the four hundred inhabitants to leave the camp and make their way safely to Janos. They could either leave or risk starvation or annihilation.

The Indians were quick to destroy the abandoned camp. They burned buildings, ruined crops and carried off anything useful. The Apaches were so pleased with their victory that they danced and celebrated for several days. They thought this meant the end of Santa Rita del Cobre.[17]

11

Guardian of the Frontier

The Mexican government could not afford to lose Santa Rita del Cobre. The mines produced exceptionally fine copper which was vital to the economy of the country. Santa Rita was the only settlement in the heart of Apachería. The Apaches would claim a significant victory if McKnight and his people did not return to the copper mines. The importance of Santa Rita del Cobre is obvious from the steps taken to secure this small, isolated camp.

Commandant General Calvo, who was also governor of Chihuahua, always gave strong support to the Santa Rita mines.[1] Chihuahua now claimed this valuable but troublesome area in the extreme northwestern corner of that state.[2] Santa Rita was important to the economy of Chihuahua, not only for the copper produced, but also for the jobs provided in many fields connected with the mining industry. It was the only active copper mine in the state, and provided that metal to the mints of Chihuahua and Mexico City for coinage. Skilled miners, common laborers, farmers, ranchers, merchants, muleteers and teamsters all benefitted from the success of the Santa Rita del Cobre mines. Candle makers, tanners, charcoal burners and others supplied items necessary for the business of mining and smelting. For thirty years, this copper mining camp served as an information center concerning the activities of Indians who roamed Apachería. It was an outpost from which to launch punitive expeditions against hostile Apaches, and the site of peace negotiations with the Indians. The Commandant General declared that Santa Rita del Cobre must be held at all cost.

Robert McKnight and Steven Courcier had some apprehensions about resuming the mining operation. The partners had made a sizable fortune from the copper mines in the past seven years, but suddenly the future of Santa Rita was uncertain.[3] The workers were reluctant to return to the mining camp after the frightening experience of being run out of their homes by the

Apaches. However, these were hard times; the whole country was suffering from a deep depression and jobs were scarce. At Santa Rita, although the work was hard and the workers existed at the poverty level, food and shelter were provided. Some of the workers, who were deeply in debt to the mining firm, had to either continue working for McKnight or go to debtor's prison. If McKnight and his people returned to Santa Rita, it was inevitable that the Indians would continue to harass the camp unless there was a large force of armed men to protect the operation.

The Commandant General envisioned a strong military installation in the middle of Apacheria, aimed at bringing the rebel Apaches under control. With this goal in mind, he appointed Captain Cayetano Justiniani[4] as commander of the 2nd Division. Justiniani departed for the copper mines with a contingent of eighty-six men from the Durango artillery and Chihuahua civil militia, and thirty-nine shackled prisoners who would serve their sentences doing public work at Santa Rita. When the 2nd Division reached Janos on November 21, 1834, twenty men of the Janos cavalry were added to the ranks.[5] At the copper mines, a forty-man company from Carrizal and twelve men from each of five other frontier presidios joined the newly formed division. The 2nd Division, now 206 men strong, was ready to carry out the orders of the Commandant General to give Robert McKnight the necessary protection to conduct his business.[6] The troops would guard the camp, escort the wagon trains, provide mail service, and take offensive action against Indians who failed to honor the terms of the latest peace treaty.

With military strength at hand, McKnight and his employees returned to the copper mines. The Apaches had ransacked the camp, taking food and livestock; they had even set fire to homes and buildings as they left the area. The extensive damage took some time to repair, but the little furnaces were again turning out copper ingots by the end of the year (1834). Courcier retained control of the copper market and set his own price. He made sure the partnership continued to prosper.

Captain Justiniani was concerned with the well-being of the soldiers under his command. Troops and prisoners were living in temporary shelters. Many of the men became ill from the cold, blustery, winter weather. Adequate quarters were needed for the soldiers, with a place to house the penal colony necessary as well. Colonel Calvo sent two engineers from Chihuahua to oversee the construction of a presidio. Between sixty and seventy-five prisoners were now available for the work force, with more convicts added to the penal colony each month.[7]

Justiniani selected a site southwest of the settlement for the fortification. He chose a high mound commanding a view of the trail that approached the camp from the canyon below. Construction followed designs used by the Spaniards for their frontier presidios, with one exception. Most of the presidios were constructed in a square or rectangular shape with towers at two opposing corners. The Santa Rita fort was built in the form of an equilateral triangle, two hundred feet to a side, with a *torreón,* or circular tower, at two of the angles. The variation may have been necessary in order to make use of this particular site. The builders used materials at hand: stone for the foundation, adobe bricks for the walls, and vigas (rafters) of sixteen to eighteen feet cut from nearby stands of tall pines. A barracks along the inside of one wall housed the officers and men. Cramped cells in which to confine the prisoners, storage facilities, a commissary, tailor shop and other everyday essentials of military life adjoined the other walls.[8]

The presidio at Janos supplied Santa Rita with two small cannons to be mounted behind the parapet to defend the fort. They also contributed a meager supply of ammunition consisting of eight cannon balls, seven small sacks of grape shot, fifteen cartridges, and one box of powder, which was all Janos could spare.[9] The Santa Rita presidio later acquired another cannon and two stone mortars (a short-barreled cannon which could be moved from one place to another). It was difficult to keep these artillery pieces in working order, and ammunition was always scarce.

Captain Justiniani encountered serious problems in providing food, clothing, arms and ammunition for the 2nd Division. All the frontier presidios were in dire need of these items, and the military budget was inadequate. Santa Rita was far from the source of supplies, which made it doubly hard for the commander to stock the commissary to feed two hundred soldiers and seventy-five convicts. The diet was simple, with tortillas, beans and a small amount of meat, with chile to make it agreeable to the taste. However, even corn for making tortillas was sometimes hard to obtain because of transportation problems. Justiniani issued a call for help. San Buenaventura sent eighteen bulls and one hundred bushels of corn. Janos furnished grain, and the alcalde offered five hundred sheep. El Paso and Sonora sent food and supplies. Courcier brought provisions in his wagon trains, and as many pesos as he could gather. Colonel Calvo provided twenty-five hundred pesos, one hundred shoes and fifty-two hoes, along with a reminder that he expected the division to plant grain near the presidio to provide for future needs.[10]

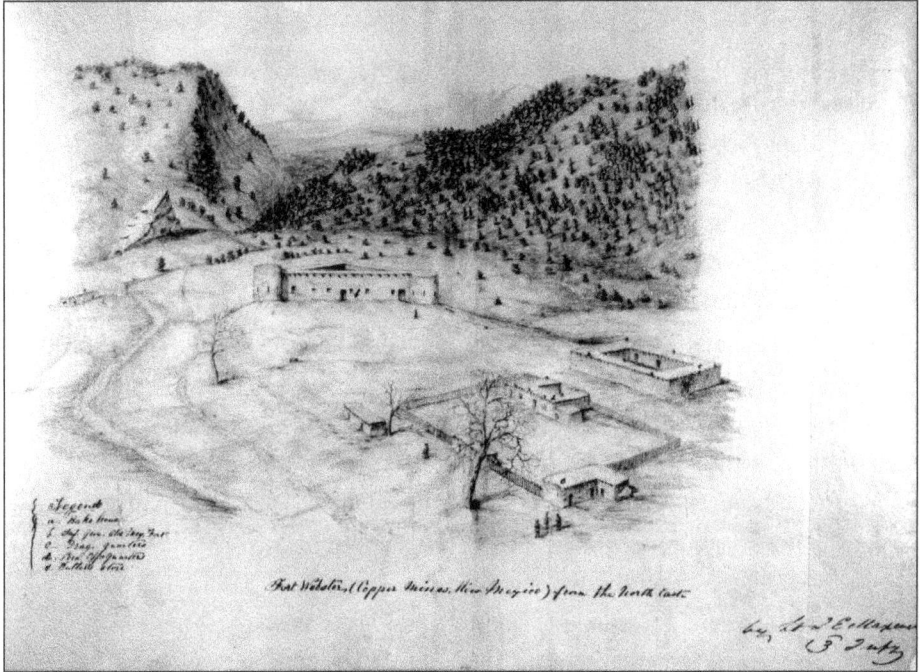

The presidio. This adobe fort was erected at Santa Rita del Cobre in 1835.
It was manned by the 2nd Division of the Mexican Army to protect the
mining operation and wage war against the Apaches. Sketch in 1851 by
Lieutenant E. Maxwell, 3rd U.S. Infantry. National Archives, Washington, DC.

The Apaches were the next problem demanding Captain Justiniani's attention. The Indians watched the flurry of activity at Santa Rita with alarm. Robert McKnight not only returned with his workers, he also brought an army with him. The Apaches lay in wait to grab the mail pouch in order to learn what the military planned. Were the soldiers preparing for a campaign in retaliation for the damage to Santa Rita? The Apaches did not wait to find out. Early in January, Juan José Compá and several chieftans appeared on the outskirts of the of the camp ready to negotiate a peace treaty. Santa Rita del Cobre had been the meeting place for peace talks between the Mexican military and the Apaches of the area for the past decade. The resulting peace treaties were fragile contracts at best. None of them led to a lasting peace, but they brought both sides a brief respite from open conflict. Mexican soldiers waged war against the Apaches to punish them for their crimes. The Indians retaliated by stepping up their plundering of the frontier settlements. The army, with military strength at a low ebb, could do little to improve the situation.

Juan José began the talks by presenting the Apaches' reasons for their wayward behavior. He said his people had experienced suffering and mistreatment at the hands of both military and civil authorities. The Indians were accused of crimes which they did not commit and punished unjustly. Many people from the settlements and presidios had abused and insulted them, killing some Indians without reason. He admitted that the Indians had caused some damage, but what else could they do to secure their own safety and survival? Now they admited the error of their ways and asked for forgiveness.[11] Justiniani, with many years of experience in dealing with the Apaches, was skeptical, especially since some of these same Apaches were responsible for forcing the abandonment of Santa Rita less than three months earlier.

Juan José was a shrewd negotiator; he wrung every possible concession from the Mexican government. The Apaches agreed to cease their raids on the presidios and settlements if the soldiers would stop harassing them. They would return all captives in exchange for tribesmen held prisoner by the Mexicans. The Indians would establish rancherías in their assigned areas and take up farming.They firmly refused, however, to return animals and goods already taken on their raids.[12] The Mexican negotiators just as firmly refused to reinstate the ration system because they could not afford the cost. Each of the sixteen chiefs who signed the peace agreement received two horses for plowing the fields and a small amount of gun powder for hunting. At Juan José's suggestion, the chiefs were granted a monthly subsistance of twenty pesos, so long as they remained at peace.

Sonoran officials refused to participate in the peace talks. They thought certain elements at the copper mines (Robert McKnight in particular) had furnished the renegades with arms and ammunition in exchange for stolen mules.[13] Too many treaties had been made and broken in the past. The Sonorans had no faith that Juan José Compá or any other chief could confidently make promises for the fiercely independent bands. Sonora's reluctance to participate weakened the agreement between Chihuahua and the Apaches and caused trouble in the future. The Apaches felt no obligation to spare Sonora from their depredations, and more than once Sonoran troops crossed the border to attack the rancherías of Chihuahua Apaches in retaliation.[14]

After three months of negotiations, sixteen Apache leaders signed an agreement which proved to be worthless. Again they choose Juan José Compá as their general. Compá had little influence with the other chiefs, but he was fluent in Spanish and skillful at the bargaining table. Justiniani rewarded the Apache general with a pair of first-quality wool pants with a stripe down the side, a dress coat trimmed with fringed epaulets, a sombrero, and a walking cane with a silver handle to emphasize the importance of his position.[15]

The Apaches were satisfied with the concessions they received. Colonel Calvo and Lieutenant Justiniani thought they had done the best they could under the circumstances. The Chihuahua citizens, whose livestock the Apaches had stolen, were angry when they learned the Indians would not return their animals. To appease them, Calvo promised that the state would compensate them for their loss if they could identify their animals. The Apaches would have their own *AB* brand in order to keep the stock now in their possession. The sixteen chiefs and their followers brought 493 horses and mules to Santa Rita to be registered and branded.[16]

The ink was barely dry on this latest treaty when the threads began to unravel. Within two months, there were reports of depredations committed in Chihuahua and Sonora. Pisago Cabezón was leading the hostiles. Although he signed the treaty in April, he had not appeared at Santa Rita since that time. The governor declared him an enemy and ordered his arrest if he appeared at any of the settlements in Chihuahua, Sonora or New Mexico. Pisago was able to gather a large following of Apaches, who caused much trouble on the frontier.[17]

The ease with which the Apaches found markets for their stolen livestock and other plunder, compounded the problem. No doubt unscrupulous foreigners, usually cited as *norteamericanos,* carried on a clandestine trade with the Indians. Mexicans were also involved in trading with the Apaches for stolen goods, but the Americans had the edge with their good firearms and fine-quality powder. Mexican officials were justified in claiming that this illicit trade encouraged Indian depredations by providing a market for stolen goods, especially horses and mules, which brought a good price in Missouri. Provided with a profitable outlet for their ill-gotten gains, the Indians increased their raids to acquire more trade goods. Even as Justiniani bargained for peace at Santa Rita del Cobre, a party of seven Americans, led by Santiago Bone (James Bonny) of San Miguel del Vado, met with Apaches at the *Picacho de Mimbres* (now called Cooke's Peak) and exchanged gunpowder for mules. Ironically, Bonny and his companions had no more than gotten home with their livestock when they were attacked by another band of Apaches who stole two hundred mules from them.[18]

The Santa Rita presidio was in serious trouble. The military supply system broke down because of the Apache raids in Chihuahua and lack of funds in the state treasury. The commissary at the Santa Rita post was depleted to the point where the possibility of starvation became a cause for alarm. Several wagon-loads of foodstuffs expected from San Buenaventure failed to arrive, and the condition of the troops grew worse each day. McKnight was able to provided the garrison with enough supplies to last a few weeks. The commander arranged to purchase grain from Janos and San Buenaventura, only to find that the grain could not be delivered to Santa Rita because Apaches had depleted the military's mule herds at those presidios. Courcier came to the rescue by sending his own mule trains to pick up whatever food and grains were available.[19] Robert McKnight shared what he had with the Santa Rita troops on several occasions when the government supply train failed to arrive. Justiniani declared that without this help, his troops "would have perished from hunger several times."[20]

Captain Justiniani found a temporary solution to his dilemma. When the forty-man company from Carrizal arrived at Santa Rita del Cobre in November of 1834, they brought only enough rations for one month. The commander wondered how he could take care of these men when no arrangements had been made for their subsistence. By March of the following year, he could no longer provide them with even the minimum amount of food necessary to sustain life. He solved the problem by sending the forty men back to Carrizal. This left the presidio with 168 soldiers to protect the mining camp and escort the wagon trains.[21]

Money was still desperately needed on the northern frontier. Colonel Calvo made numerous appeals to the Mexican Congress and to President Santa Anna concerning the general scarcity of supplies and lack of funds with which to carry out the war against the Apaches. The central government was concerned with rumbles of rebellion in Texas at this time. American settlers had been allowed to take up land in that state. They now outnumbered the Mexicans living there and were agitating for independence from Mexico. Troops and supplies were needed in that quarter. Mexico was also at war with the Comanches who were causing much trouble in Texas and eastern New Mexico. Nevertheless, President Santa Anna did what he could to meet Colonel Calvo's needs. Supplies brought into the country at the port of Matamoros, on the Texas coast, were exempt from duty up to the amount of fifteen thousand pesos per month. Calvo also received a draft for twelve thousand pesos, along with two hundred rifles, two hundred muskets and a box of medical supplies.[22] All of the presidios under Colonel Calvo's command needed help. The government aid, divided between several needy presidios, did little to improve the situation.

The troop strength at Santa Rita del Cobre was now reported at one hundred fifty-four men. Of this number, an officer and twenty-three men guarded the remount herd. It was necessary to pasture these animals at Pachitejú or in the Mimbres valley, some distance from the mining camp, to provide water and good grazing. At both places they were vulnerable to Apache thievery. A corporal and fourteen men escorted the wagons and two men carried the correspondences between Santa Rita del Cobre, El Paso del Norte and Janos once a week. Two or three soldiers always guarded the prisoners when they went to gather firewood or performed other assigned tasks. Guard duty rotated weekly among the troops. Forty or fifty men went out regularly on patrol to search for hostile Indians and stolen animals. When off-duty, the soldiers had mini-target practice; each man fired only two shots because of the scarcity of ammunition.[23]

The entire frontier presidial system continued to suffer from lack of funds. The soldiers were often on short rations, clad in tattered uniforms and armed with inferior weapons. The men were untrained and poorly paid. Salaries were usually several months in arrears, and families suffered from privation. This led to discontent most of the time and occasionally downright rebellion. When Colonel Calvo ordered Justiniani to prepare for a general campaign against the hostiles using the combined forces of the frontier presidios, Justiniani replied that an effective campaign would be difficult, if not impossible, because the presidial companies were in such poor condition that it was hard for them to carry out their regular duties.[24]

For example, Lieutenant Mariano Rey, commander of the presidio at Janos, was ordered to furnish an escort to protect the convoys of provisions going to Santa Rita and to establish a bi-weekly mail service to that place. He explained that the Apaches had run off most of the presidial horses and there were not enough animals left to mount five men. Calvo assured him that the stolen horses would be replaced as soon as possible. In the meantime, the lieutenant would carry out orders, even though it meant doing so on foot. The Janos garrison was still without horses several months later. Lieutenant Rey reminded the military commander that due to the lack of mounts, the mail and escort services were still carried out on foot. With the speed of the wagons and the long distance they traveled between Santa Rita and Janos, the soldiers returned exhausted and emaciated, with swollen and blistered feet. They always needed several days of rest before the next trip. The Janos company usually provided twelve to fifteen soldiers to guard the wagon or mule trains. Lieutenant Rey complained that he was running out of men for this duty.[25]

Soldiers continued to desert the army and flee from the Santa Rita presidio in alarming numbers. Many wanted to join the fight brewing with Texas rebels. In an attempt to inspire the soldiers with a sense of responsibility and loyalty, General Calvo sent a stirring message to the troops at Santa Rita to assure them that he was well aware of the misery and suffering they were enduring and would make every effort to find relief for them. Once again he stressed the need to defend the mining camp because it was the center of operations against the Apaches, and "is as important as Texas to the republic. By maintaining a strong force in the heart of enemy country, Apache incursions into Chihuahua and Sonora occurred less frequently," he said. " Furthermore, if Santa Rita is abandoned, the Apaches will count it as a great victory and be quick to celebrate." In conclusion, Colonel Calvo reiterated the significance of Santa Rita del Cobre as "it should be looked upon as the defense of the western frontier and the *atalaya* (watchtower) guarding the families and interests of our fellow countrymen."[26]

12

Camaraderie and Chaos

Santa Rita del Cobre had a population of between five and six hundred people by the end of 1835. These simple, hard-working folk deserve recognition in the story of Santa Rita. Their names seldom appear in the records unless they commited a crime, yet they were as important to the success of the copper mines as the administrators, military commanders, and government officials. José E. Espinosa described them well in *Saints in the Valleys*: ". . . the people of that time and place were frontiersmen with all the moral virtues and weaknesses, all the practical gifts and intellectual shortcomings of any group living on the fringe of civilization.[1]"

Life was hard for both miners and soldiers at the isolated camp. Hostile Apaches were only part of the problem. Living conditions were primitive and barely above the poverty level. The mine workers usually built their own shelters, either crude wooden jacales or one-room adobe huts, scattered at random around the small valley. The only furniture was homemade and rustic. At night they rolled themselves in a blanket or serape, and slept on a pallet on the hard-packed dirt floor. On the positive side, however, McKnight and Courcier took care of their workers. They received their pay on time, along with their ration of corn, beans and meat. McKnight's company store offered a variety of goods to enhance their lives. With every wagon train, Courcier sent food and dry goods for the store. Although the wares offered were not as plentiful as in previous years, the inventory included such popular items as soap, cakes of crudely refined sugar called *panocha* and chocolate. The workers needed citrus fruits and small quantities of various medicinal plants to maintain their health. Cotton and woolen textiles for clothing were much in demand. What Courcier could not obtain in Mexico, he imported through the Texas port at Matamoros.[2] John McKnight also brought merchandise for the mine from the United States over the Santa Fe Trail. Employees could buy on credit or receive part of their wages in merchandise. If the worker committed a serious crime, McKnight turned him over to the penal colony to serve his sentence. They left no record of dissatisfaction or desertion.

The soldiers, on the other hand, lived in a half-finished presidio, with interior doors and windows or shutters missing. They wore ragged uniforms, and seldom received their pay regularly. Weapons were in poor condition and gun powder was so scarce that each man could only fire two shots at target practice. The cavalry complained because they had no horses, while the infantry bemoaned the fact that they had no shoes. Starvation was a serious threat for the military and penal colony when government shipments of food and supplies did not arrive. With the military on reduced rations, morale was low, prisoners escaped, and soldiers deserted.

Even in these troubled times, the soldiers and miners and their families found time for social activities. They were happy, exuberant people, who enjoyed camaraderie and celebrations. Although Santa Rita had no church or resident priest, the faithful celebrated religious holidays and held fiestas to honor patron saints. Saint Rita of Cascia, the saint of the impossible and the advocate of desperate cases, received special reverence as the guardian of the mining camp. Horse races, foot races, and cockfights relieved the monotony of hard work and daily routine. Someone always had a guitar or fiddle to provide music for singing and dancing. Everyone loved the *bailes*, where they danced their favorite fandango through the evening hours. The authorities frowned upon gambling, card games, and heavy drinking at the mining camps because these diversions often ended with hot tempers and brawls. Nevertheless, covert games of chance, monte, and dice were major amusements.[3] Santa Rita probably had a few cantinas (saloons) serving wine and brandy brought in from El Paso del Norte. Mescal, a potent home-brew made from the fermented juice of the agave (century plant), generally enlivened the festivities.

With such a hodgepodge of people, confined by circumstances into a limited space and living in constant danger of Apache raids, trouble was bound to come along. Soldiers and miners disagreed from time to time. The scarcity of women in the mining camp often caused trouble. For example, Felipe Borsigo, a worker in the mines, claimed Guadalupe Molina as his woman. Guadalupe also shared her favors with Private Santiago Zapata. One night, Felipe arrived at the home of Guadalupe for a visit and found Santiago there. When Guadalupe tried to leave, Santiago struck her, knocking her to the floor. A fierce argument followed. Felipe went after Santiago with a stick, and Santiago stabbed the miner in the shoulder with his lance, causing a serious wound. Felipe picked up the soldier and threw him into the arroyo. Guards arrested both men and took them before the commander. Justiniani found Private Zapata guilty, although both parties agreed that the deed was done partly in self-defense. The military commander sentenced Zapata to serve two months in the guard house.[4]

Short rations, delinquent pay, and miserable conditions led four soldiers to desert, taking arms, ammunition, and horses with them. When one young private fled Santa Rita, he took not only musket, sword, and horse, but also María, the wife of a mine worker. Military authorities arrested the soldier and the errant wife when they reached Janos. The soldier returned to Santa Rita as a prisoner to serve his sentence in the penal colony. The fate of the wife is unknown.[5]

Another soldier shot himself in the hand while guarding the horses. His commanding officer suspected a self-inflicted wound. Did the soldier think this would earn him a discharge so that he could return to Janos where his new bride awaited him? Justiniani held an investigation, but the soldier swore that his gun had accidently discharged when it got caught in his serape. Justiniani decided to consider the shooting an accident. Because he needed every man at the presidio, the commander ordered the wounded soldier to return to duty.[6]

Rivalry between the soldiers of the regular army and the men serving in the civil militia created problems now and then as well. The presidial officers were scornful of the militiamen, comparing them to a semi-armed mob rather than an army. They lacked discipline, uniforms, and rifles, and had little training. The regular forces said they could not depend on the militia in battle. This was hardly fair to the militiamen who were forced to serve when called upon. They furnished their own supplies, horses, equipment and weapons. They fought on their own time, at their own expense, with very little government support.[7] One petty feud that threatened to become serious was the case of Lieutenant Inocencio Cisneros, of the Chihuahua militia, who was in charge of the penal colony. Lieutenant Hernández, a cavalryman in the regular army, spread a rumor that Cisneros planned to desert and join the Apaches. Cisneros felt this preposterous story was a slur on his honor, and went looking for the cavalryman, ready to demand satisfaction. The commander had both men arrested and brought before him. The cavalry officer apologized to Cisneros for the "misunderstanding," declaring that everything he said was in jest. Tempers cooled, and the commander dismissed the case. This was just one among many such incidents between soldiers of the regular army and the civil militia.[8]

The arrival of Captain William Dryden, of the Chihuahua militia, only added to the chaos at the mining camp. Colonel Calvo decided the state should have a share in the Santa Rita mining business that was so lucrative for McKnight and Courcier. He hoped to defray some expenses of the war against the Apaches with profits from a copper mine. Calvo appointed William Dryden, an American living in Chihuahua, to carry out his plan. Dryden, a native of Kentucky, came west as a Santa Fe trader in 1827. He established his headquarters as a merchant in Chihuahua City and served as a captain in the state militia. Colonel Calvo directed Captain Dryden to march to Santa Rita, escorted by forty men of the Chihuahua militia assigned to the 2nd Division. There he was to locate a good vein of copper, register it in the name of the state mint, and begin production as quickly as possible. Convicts serving time at Santa Rita would work in the mine. Calvo told Dryden to organize a local militia company to protect the state business and escort the copper shipments to the mint.[9]

William Dryden was a rascal, and he brought nothing but trouble to Santa Rita del Cobre. Colonel Calvo authorized Dryden to arrange for anything he needed for the success of his project. Evidently, Dryden thought this gave him full authority at the mining camp. First he challenged Captain Justiniani over command of the presidial troops. Justiniani quickly put him straight on who was the military authority at the Santa Rita post. Dryden met the same resistance when he tried to take over the local civil militia made up of mine employees. The miners refused to obey anyone but McKnight. Colonel Calvo suggested that Justiniani form a militia company using vagrants and malcontents to guard the state mining business. Justiniani replied that he could not comply with this order because McKnight permitted no individuals of this nature in the camp.[10]

Dryden lacked adequate mining experience to develop the successful copper mine that Colonel Calvo envisioned. Having already antagonized Captain Justiniani and Robert McKnight, he received little help from them. The Santa Rita area had an abundance of copper, however, and Dryden had no trouble finding a vein that looked good. He put seventeen convicts to work developing the claim. They would receive, from the profits of the mine, one real (about twelve and one-half cents) per day for a single man and one and a half reales for married men.[11] The director of the Chihuahua mint arranged for McKnight to refine the ore into copper bars which Dryden would then transport to Chihuahua in state-owned wagons. The anticipated profits from this copper mine would greatly improve conditions at the presidios and build a stronger military force for the war against the Apaches.

An added burden for Captain Justiniani was the administration of the penal colony under his command. During the first year, Chihuahua sent more than sixty criminals to Santa Rita. Some arrived without papers, leaving Justiniani without record of the crimes committed or length of their sentences. He had to provide rations of corn, beans, and a small amount of meat for them, adding to the strain on the slim stores of the presidio commissary. The situation became desperate at times, with little hope for improvement. The prisoners worked hard during the day; at night they were confined in small windowless cells. After several months of incarceration, Francisco Orcacitas petitioned Justiniani to allow him to move about within a small area outside his cell. He needed fresh air and conversation with fellow prisoners because the conditions of his imprisonment had placed his health in a deplorable state.[12]

By the end of the first year, one-third of the convicts had escaped, choosing to risk capture or death from the Apaches rather than stay in such a miserable place. They often took a horse or two, and whatever supplies they could lay their hands on. The prisoners escaped so frequently that Justiniani issued orders that any guard who let his charges get away would be severely disciplined. Rather than pay the penalty, the guards often deserted and joined the fleeing convicts themselves. Lieutenant Inosencio Cisneros, the officer in charge of guarding the prisoners, asked Justiniani to send him on a campaign against the Indians rather than deal with such an impossible situation.[13]

Health problems were common on the frontier, with few doctors and little in the way of medicine to deal with ailments. The people relied on tonics, elixirs, salves, and herbal potions to care for the sick or injured. Each monthly military report of troop strength listed several soldiers too ill to carry out their duties. Besides common ailments such as colds, influenza and stomach disorders, Santa Ritans had to deal with battle wounds, mine accidents, broken bones and epidemics of contagious diseases. Smallpox was the scourge of the frontier. It was especially hard on the Apache population if they contacted the disease. In a report of 1834, the Apaches claimed that one of their men became ill a few days after he shared a cup of coffee with Americans. The infected Apache carried the disease (probably smallpox,) to other Apache rancherías. The Indians claimed that more than 300 Apaches died of this epidemic.[14]

Hilarión García, a merchant from Sonora, had a store at the copper mines, and charged less for his merchandise than the price of goods at the company store. Soldiers could trade at Garcia's store, but the military sometimes did not receive any pay at all for three or four months at a time, and had no money to spend. Itinerant peddlers from Chihuahua and Sonora visited the camp regularly, bringing in merchandise surreptitiously to avoid paying the sales tax. The government appointed Robert McKnight as tax collector to put a stop to the smuggling.[15]

The Mexican government controlled the tobacco industry. Robert McKnight was also *fiel de tobacos*, the agent who received shipments of government tobacco, sold the products at the company store, and sent the revenues to the treasury in Chihuahua. Calvo ordered McKnight to turn over all revenues from the sale of tobacco to Captain Justiniani to use in purchasing supplies for the 2nd Division. The amount McKnight collected in six months totaled more than two thousand pesos. Justiniani declared that without this help, his troops would have perished from hunger.[16]

Because tobacco was under government control, citizens could not grow the crop for their own use. Smoking was a common past time for men, women and even children as young as ten years. Many people could not afford to pay the price of government tobacco, but *punche* was always available. This was a native tobacco of poor quality that grew well at high altitudes. Although the government prohibited its cultivation as a violation of the tobacco monopoly, folks continued to raise punche in secluded places. When the plant matured, the leaves were dried, crushed with the fingers, and rolled in a square of corn husk for cigarettes. They rolled whole leaves tightly to form a cigar. Punche was cultivated in isolated locations throughout the frontier and traded or sold on the black market. Mexican officials suspected the illegal cultivation of punche at Santa Rita. Colonel Calvo ordered McKnight to destroy all crops of tobacco found growing in the area and prosecute those responsible. As this would cause a hardship for his workers, McKnight probably found no illegal crops of punche.[17]

Colonel Calvo planned a general reorganization of the forces early in 1836, hoping to strengthen the frontier line of defense. He ordered Captain Justiniani to the capital to resume his duties as secretary to the Commandant General. Calvo appointed Captain Mariano Ponce de León of San Buenaventura commander of the armies of the western frontier to take Justiniani's place at Santa Rita. Lieutenant Mariano Rodríguez Rey, of the Janos cavalry company, was second in command. Colonel Calvo ordered a company of soldiers from the capital to man the presidio at Carrizal; the 4th Active Company of Carrizal would move to San Buenaventura. San Buenaventura Company would move to Janos, where the remaining thirty-five men of the Janos Cavalry would join the rest of their company at Santa Rita del Cobre.

The ninety-six men of the Janos Cavalry would form the backbone of the forces at the Santa Rita presidio. Many soldiers in this company had served at the copper mines since the fall of 1834. Their families remained in Janos. With the head of the household absent, a family could never be sure of getting its share of the food and clothing rationed out by the Janos supply officer. This led to complaints and desertions. Colonel Calvo decreed that when the remainder of the Janos Company moved to the copper mines, the families would also move there. He advised Captain Ponce de León to use convict labor to build houses outside the fort but close enough to protect the families.[18] Robert McKnight suggested that the Commandant General appoint a primary teacher to instruct the children of the soldiers and miners. Calvo, a strong supporter of education, quickly approved McKnight's suggestion. The school could meet in a small house originally built for Juan José Compá, but never occupied. Calvo advised McKnight to go to Villa del Paso to find a person qualified for the job. McKnight's success or failure in founding a school at the mining camp, however, is not recorded.[19]

Lieutenant Rey felt that bringing wives and children to Santa Rita would only create more chaos. He warned the Commandant General about the chronic shortage of food. It was imperative that the division supply officer received additional corn, meat, and other rations necessary for their sustenance before the Janos troops and families arrived. Otherwise, they would starve. In spite of Rey's objections, in April of 1836 the remainder of the Janos Cavalry Company transferred to Santa Rita, bringing all the families who were willing to make the move. Father Rafael Echeverría, the company chaplain, also transferred to the copper mines. For the first time, Santa Rita had a resident priest. In the next six months, he baptized twenty children, performed one marriage and directed three funerals.[20]

As Lieutenant Rey predicted, the addition of families to the military colony placed a severe strain on the already scarce food supplies at the presidio. Captain Ponce de León could not get corn and grain from either San Buenaventure or Janos because again the Apaches had stolen all the mules from both presidios. Ponce de León informed Calvo that the small amount of meat and corn on hand was almost gone. McKnight and Courcier always did what they could to help the military installation at Santa Rita. This time, however, they only had enough foodstuffs to take care of the mine employees and their dependents. The soldiers threatened to leave if something was not done to relieve their hunger. The captain refused to be responsible for the consequences if conditions at the presidio did not improve.

Colonel Calvo was aware of the plight of the people living at Santa Rita del Cobre, and knew that if Santa Rita was to survive, the persistent problem demanded a solution. The government of Chihuahua was in "anguished circumstances" itself, what with confronting Indians ready to ravish the frontier and Americans in Texas who were about to take the Republic apart.[21] The state could furnish only a small amount of supplies. Calvo notified Colonel Conde, the commander at El Paso del Norte, of the urgent needs of the Santa Rita presidio. The El Paso commander dispatched a company of civil militia with two hundred fifty bushels of grain for the immediate relief of the isolated garrison. He ordered the militia to take over the Santa Rita presidio if they found it already abandoned.[22]

Colonel Calvo established the presidio at Santa Rita del Cobre, manned by an entire military division, to act as "guardian of the western frontier." He believed that respect and fear of the troops in the heart of Apachería would protect the mining camp. The soldiers would maintain some measure of control over the Apaches, and curtail their frequent incursions into Chihuahua and Sonora. It was an idealistic theory that never came into being because the military could not sustain the appearance of strength and authority, which they no longer possessed. It was obvious to the Apaches that they could unite and gain control of Apachería.

13

Fire in the Hole

Miners of the early days, when blasting to break out an ore vein, would load a hole or crevice with a charge of gunpowder and secure a fuse. When all was ready, the blaster shouted "fire in the hole" to warn any workers nearby to make a rapid exit to safety before he lit the fuse.[1] Santa Rita del Cobre, besieged by trouble and misfortune, was like a charge of black powder waiting for some calamitous event to light its fuse.

The Apaches took to the warpath with renewed confidence and vigor in the spring of 1836. Warriors and their families gathered by the hundreds along the San Francisco River.[2] Apache chiefs Pisago Cabezón, Fuerte, Mano Mocha, Chirimi, Juan José Compá and his brother Juan Diego, and several other leaders were there with their bands. They were now united in a common purpose—to make war on the Mexican and American intruders. Although fervently independent, the chiefs and their warriors quickly learned the advantage of combining operations with other bands. Apparently, they were willing to follow Pisago Cabezón as he led the chiefs in planning a strategy for raiding and warfare. The Apaches were well armed, mounted on good horses, and both warriors and steeds were skilled at hit-and-run tactics. They presented a force which created apprehension across the frontier.

Soldiers and civilians of the Santa Rita mining camp lived in a state of suspense as they anxiously awaited the safe arrival of wagon trains bringing desperately needed food and supplies. Were the twenty armed men escorting the wagons enough to thwart the Apaches in their plan to cut off supplies from the mining camp? Would the hostile and unpredictable Apaches attack Santa Rita en masse as they had threatened? The inhabitants knew they would soon have their answer, as the situation became more and more explosive.

In April of 1836, Apaches from a camp in the Mimbres valley sent word to Santa Rita that they wanted to begin peace negotiations. The next day four chiefs, accompanied by several braves and women, came to the mining camp.

After a day of palaver, the Apaches sent word to Pisago Cabazón, Juan José Compá, and Chirimi, principal leaders of the present hostiles, to join them in making a peace treaty. Lieutenant Rey identified the Apaches as members of a large group that had recently attacked Sabino (Bernalillo County). They killed three herders, took several captives and stole more than four thousand sheep. The brazen thieves then tried to sell the stolen sheep to the inhabitants surrounding Sabino. Rey informed Colonel Calvo of this latest development, ending his report with this caustic comment: "The seeming desire for peace stated by these Indians could possibly be true, as they are very disappointed in not having anything left to steal."[3]

A few days later, Pisago and Juan José appeared at Santa Rita. Pisago offered to go to the Mogollón, San Francisco, and Gila areas where many bands had rancherías, to see if they were interested in making peace. Lieutenant Rey thought Pisago Cabazón would be the next "General," as Juan José had lost much of his influence over the tribes. In this instance, Pisago was almost too successful.[4]

The residents of Santa Rita thought the end had come, as Apaches arrived in increasing numbers to the surrounding hills. It looked as if all of Apachería was there. About midday, a group of mounted and well-armed warriors, rigged up for war, appeared on a hill near the mine. They made a big commotion, which got everyone's attention. All that kept the Indians from attacking was the cannon always in readiness and the one hundred twenty-two soldiers on duty. The enemy advanced rapidly in considerable numbers and took over all the roads leading out of the camp. The inhabitants of both mining camp and presidio wondered if this multitude were here to talk peace or to make war.[5]

Juan Diego Compá came in to demand the release of two Apaches held prisoner at the presidio. Lieutenant Rey offered to release the Indians in exchange for two of the "most wanted" warriors, one of whom they knew had stolen all McKnight's mules. Juan Diego declared that the Apaches wanted peace, but nothing was gained by a discussion which lasted all day. The Indians slaughtered a few cows and withdrew to the nearby mountains for the night, only to reappear in force the next morning.

The inhabitants of Santa Rita del Cobre were very apprehensive; no one knew what to expect from the horde of Apaches surrounding the camp. Lieutenant Rey advised Juan Diego to warn his braves that the soldiers would fire on them if they came too near the settlement. Juan Diego took this message to the waiting Apaches and then returned to McKnight's house to continue the talks. A short time later, the situation became precarious when seven defiant Apache braves started down the hill toward the mining camp. Juan Diego quickly went out to stop them, while Lieutenant Rey ordered the troops on the roof to hold their fire. No one obeyed. The Indians continued to insult the troops, and the soldiers discharged their stone mortar at them. All the Indians, including Juan Diego, quickly disappeared. This ended the peace negotiations.[6]

By this time word of the critical state of affairs reached the Commandant General, who ordered a large force of frontier troops to go to the assistance of Santa Rita. The expected attack from the hostile Apaches never came. Patrols went out each day, but they found no sign of the Indians. Two weeks later, the Apaches returned to Santa Rita in full force and began to make extensive preparations for war. Their smoke signals appeared daily. They brought in cattle and horses and left them in different places around the hills to provide food for the warriors. Lieutenant Rey learned that the Indians planned to attack the mining camp for six days and nights from all directions. The Apaches only waited for the Coyoteros and Navajos to join them and assure the success of their undertaking.[7]

Apache allies failed to arrive. The Santa Rita del Cobre presidio, temporarily reinforced by troops from Janos and San Buenaventura, was now too well defended to be an easy victory. Juan José Compá met again with Lieutenant Rey and Robert McKnight. Apparently, they reached some agreement in this discussion. The Indians withdrew the following day and headed south to resume their assault on the interior of Chihuahua. Again, Santa Rita del Cobre gained a reprieve. The Apaches caused little damage other than to the horse herd. When Rey first turned the horses out to pasture at the suggestion of Juan Diego Compá, there were sixty-six animals. By the third day, the patrol could find only thirty-five horses. As a final gesture of defiance before leaving the area, the Apaches set fire to all the grazing lands around the mining camp.[8] Calvo ordered the few animals belonging to the presidio kept in a corral and fed well so they might become strong enough to withstand the harshness of frontier service.[9] Finding replacements anywhere on the frontier was hard. What good was a cavalry without horses?

The settlement at Santa Rita del Cobre had little chance of withstanding an attack from a multitude of hostile Apaches. Fortunately, the Apaches decided to seek other targets at this time. The military reinforcements returned to San Buenaventura and Janos, leaving the 2nd Division with only one hundred two soldiers. Santa Rita presidio and the troops who manned it were still in miserable condition. The exterior walls of the fort were sturdy, but the interior work was never finished. Workers had installed only three doors, and all the interior windows were missing. Furniture, if they had any, was handmade and very rustic. Gunpowder was scarce, arms were in poor repair, and equipment was missing. A four-pound cannon remained at the ready, but the gunners had only fourteen loads of round ball and thirteen loads of grapeshot to fire from it. When General Calvo ordered this artillery piece returned to the capital, the desperate Santa Rita commander managed to ignore the order. Small cannons, mounted at the port holes of the towers, were unstable or in disrepair. Firearms were lost, stolen, sold or damaged. A gunsmith came from Janos to make repairs, but he lacked the proper tools and fittings to put the firearms back into service.

The soldiers and their families went on reduced rations when government shipments of food and supplies did not arrive. The men received no pay for several months. They also needed clothing, especially pants, since a deserting soldier had taken with him a pack mule loaded with seventy pairs of trousers from the division supplies.[10] Morale was low, and the garrison at Santa Rita del Cobre was plagued with insubordination and desertion. Officers and sergeants of the 2nd Division demanded relief for themselves and their men. Captain Ponce de León, the division commander, knew their complaints were justified. Although the company already owed Robert McKnight for two hundred fanegas of corn, the commander found it necessary to turn to him once again for assistance. As usual, McKnight came to the rescue. Ponce said, "This generous man, as always, gladly granted my request. For the present I am relieved of my problem, which is to keep the troops I have the honor of commanding from dying of hunger."[11]

McKnight and Courcier also had their problems. Sometimes, the profit from the copper produced at Santa Rita did not outweigh the difficulties encountered in producing it. Mexico was not an industrialized nation, therefore, Courcier imported manufactured goods from the United States or Europe. He had to import a large variety of items vital to the mining industry, such as bars of iron and steel, wire cable, metal tools, parts, fittings and equipment. This was costly, because there was a steep import duty to pay at the custom house. To offset this expense, Courcier established a market for copper in the United States and Europe. Large freight wagons, loaded with copper ingots, made the long haul from the Santa Rita to the port of Matamoros, where he exchanged the copper for manufactured goods and processed foods from overseas. Meanwhile, McKnight was dealing with serious problems at Santa Rita. As mining operations went deeper underground, the mines began to flood at a rate that was impossible to handle by simple bailing. Carbon monoxide was a threat to the health of the miners, and cave-ins were frequent at lower levels. Sometime during this period, there was a serious cave-in at the principal mine. Thirteen miners were killed and their bodies were never recovered. The Mexican miners became superstitious and refused to go near the area. There was extensive damage to the mine, obliterating the entrance into the hillside. Only the huge waste dump indicated the location.[12] McKnight shifted operations to one of the other mines included in the lease. The partners hesitated to invest a large amount of money in developing other mines when the future of Santa Rita was in doubt. Everyone felt uneasy because of the erratic behavior of the Apaches. They came in friendship to trade one day and threatened war on the next.

Stephen Courcier still controlled the Chihuahua copper market. He received sixty-five pesos per hundred pounds for all the copper he produced. He was now in danger of losing one of his primary markets, the mint in Mexico City. The English firm of Manning and Marshall, doing business in Mexico, proposed to import three million pounds of laminated copper. They offered to sell this foreign copper to the Mexico City mint for fifty-five pesos per hundred pounds.

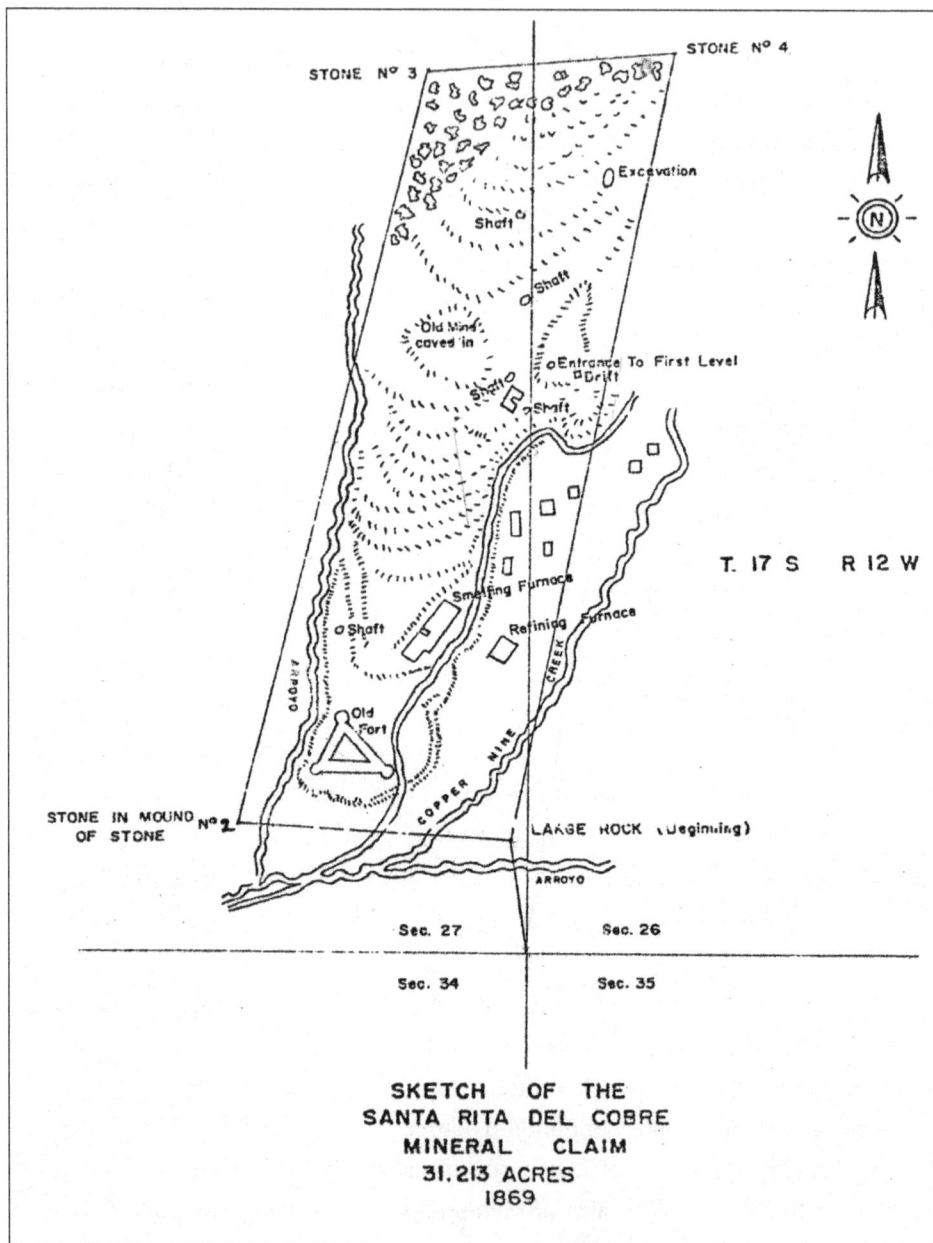

STONE N° 3
STONE N° 4
Excavation
Shaft
Shaft
Old Mine caved in
Entrance To First Level
Drift
Shaft
Shaft
Shaft
T. 17 S   R 12 W
Smelting Furnace
Refining Furnace
Shaft
CREEK
Old Fort
STONE IN MOUND N° 2
OF STONE
COPPER MINE
LARGE ROCK (Beginning)
ARROYO
Sec. 27
Sec. 26
Sec. 34
Sec. 35

SKETCH   OF   THE
SANTA RITA DEL COBRE
MINERAL   CLAIM
31. 213 ACRES
1869

Santa Rita Mining claims. This map of Santa Rita mining claims in 1869 shows the
location of early mine workings, waste dumps and buildings from previous operations.
Courtesy of Bob Eveleth, New Mexico Bureau of Mines and Mineral Resources.

Courcier and others who were interested in the copper industry wrote letters to the governor and to the editors of the local newspaper protesting the unfairness of this proposal. Courcier stressed the damaging effect bringing foreign copper in would have on Mexico's mining industry, and subsequent effects on other industries. He suggested that the government could help the existing copper miners by promoting the exploration of new mines. Idle copper mines, if reactivated, would produce enough copper for the entire nation. Courcier claimed that his Santa Rita del Cobre mines alone could furnish all the copper that the Mexico City mint needed. He reminded the governor of the military significance of Santa Rita del Cobre, stating, "The establishment at Santa Rita, supported by the copper mines, forms a frontier against the barbaric tribes."[13]

Rafael Tercero, a strong supporter of the copper industry, was very dramatic in expressing his indignation:
"It is not enough that we dress with foreign cloth and linen. Now we want to import metal to pay for our necessities. This is truly too much! We might as well lay down naked in the hot sand and give up. With the pretext of importing thirty thousand quintales of copper, they will bring in sixty thousand. If our pots are made from foreign copper, they should at least send them full of steak, roast beef and plum pudding."[14]

The military division at Santa Rita del Cobre was down to seventy-one men by October of 1836, when an incident occurred which would have grave consequences for the mining camp. The trouble began when an Apache chief known as Tapilá came to Santa Rita to trade hides and other articles for corn. He turned in a captive, José Peña, from a village near Acoma recently raided by Tapilá's band. A few days later, José Peña disappeared from Santa Rita and made his way to Tapilá's ranchería. In the dark of night, he stole sixteen horses and three mules from the Apaches and drove them at breakneck speed to the mining camp. José claimed that he stole Tapilá's horses as revenge for the damage Apaches did to his village. When Tapilá came to claim his animals, Captain Ponce de León would not release them because most of the animals had Chihuahua or Sonora brands.[15]

Tapilá was a member of Pisago Cabezón's large extended family. He returned to Santa Rita two weeks later, reinforced by Pisago and two of his sons, to request his horses. Captain Ponce de León again denied his request. After a vigorous discussion, Pisago agreed to return in a few days to discuss a peace treaty. The Apache chief promised to contact Fuerte and others of his territory to persuade them to return with him. In return, Pisago asked Ponce to grant him permission to sell Santa Rita residents a few cows and young bulls acquired on the last raid in Sonora. This put Captain Ponce on the spot. He knew the importance of winning Pisago's good will, but dealing in stolen goods was illegal. He wanted to further good relations with this important Apache leader, but was concerned about condoning the bartering of stolen livestock. In this instance, Ponce was willing to bend the rules a bit.[16]

Pisago sent three Apache men and five women to the settlement, on October third, with fifteen steers to trade. Two of the women were wives of Chief Pisago and two of the men were his sons-in-law. The trading took place with no problem, until the miners learned that one of these Apaches had previously killed a resident of the camp. They became angry and attacked the visiting Apaches. Before Captain Ponce and the company chaplain could restore order and get the Indians into the protection of the fort, two men and one woman were "beaten, stabbed, speared and shot to death." The other Indians fled, leaving the camp in an uproar.[17]

Captain Ponce ordered the two Mexican civilians responsible for inciting the riot taken into custody. Soldiers joined civilians in raising such a cry of protest that Ponce ordered the men released. Everyone knew Pisago Cabezón and his allies would seek revenge for this treacherous act. What form retaliation would take and when it would occur was anyone's guess. This ill-conceived act of violence lit the slow fuse that would end the first period of development and prosperity at Santa Rita del Cobre.

14

The Last Days at Santa Rita del Cobre

Several factors contributed to the end of mining operations at Santa Rita del Cobre. First, the copper market took a tumble when the Mexican government forced Governor Calvo to lower the value of Chihuahua's copper money by one-half. By this time, Robert McKnight and Steven Courcier had acquired mining interests in other places and devoted less time and effort to the Santa Rita mines. The ultimate cause for abandonment, however, was the inability of the presidial troops and civil militia to deal with the Apaches. The Indians now intensified their efforts to dislodge the inhabitants from Santa Rita del Cobre.

After Santa Rita citizens attacked Pisago Cabezón's trading party, there was no hope for peace. Pisago surprised and alarmed everyone by appearing a few weeks later with a small band of lesser chiefs and warriors. Pisago would not enter the camp, but he met with Captain Ponce de León on a nearby hill. The captain could do little to justify the unwarranted attack on Pisago's people, but the old chief agreed to return later and negotiate peace. Instead of appearing at the stipulated time to sign a treaty, the Apaches carried out another raid on the Santa Rita livestock.[1]

Colonel Calvo sent Lieutenant Colonel José María Arce, adjutant inspector for the state of Chihuahua, to Santa Rita to investigate conditions at the mining camp and restore order. After assessing the situation, Arce concluded that a regular supply of food and grain would do much to end the discontent of the Santa Rita citizens and soldiers. First, he authorized Pablo Moreno of Chihuahua to sell eight hundred fanegas (about thirteen hundred bushels) of corn to the Santa Rita presidio. Moreno agreed to deliver one hundred ninety fanegas of grain every two months. Fifteen soldiers would escort his supply train along the hazardous road through Apache lands. Moreno offered to provide horses for the escort if necessary. Lieutenant Colonel Arce also contacted Hilarión García of Sonora, who once had a store at Santa Rita. Garcia agreed to immediately furnish the presidio with seven loads of flour, and a load of salt. Over the next two months, he would deliver two hundred arrobas (five thousand pounds) of meat and three loads of flour. García asked Arce to pay him in silver, because the copper coins in circulation in Chihuahua were not legal in Sonora.[2]

With food on the way, Lieutenant Colonel Arce turned his attention to the state copper mine. William Dryden and the state mine were a bitter disappointment to Colonel Calvo. The Commandant General had expected profits from this mine to improve the frontier presidios and finance further military operations against the Apaches. It was not to be. By the end of the first year, Dryden had spent more than 2,500 pesos of state funds on a project which produced very little copper. He finally sent two wagon loads of refined copper to the state mint. The wagons arrived empty in Chihuahua, but no one could account for the missing copper. By the time Lieutenant Colonel Arce inspected the mine, it was flooded in the lower levels. Colonel Calvo ordered Dryden to appear before the state officials to explain the failure of the mine. Was it poor management, incompetence, a low grade of ore, or just bad luck? Dryden's excuses were unacceptable. The property needed a healthy infusion of cash to put it back into operation. State officials considered this a poor investment. They dismissed Dryden and abandoned the mine.[3]

Inspector Arce thought he had a partial solution to the chronic problem of supplying food to Santa Rita. Farmers had tried to raise crops in the Mimbres River valley several times over the past thirty years. In times of peace, they lived and worked as neighbors with the Apaches. When the peace was broken, they abandoned their land and fled to the protection of the copper mines. Colonel Arce proposed sending all the convicts to build a fortification in the Mimbres valley. They would cut fifty or sixty tall, sturdy timbers and erect a substantial palisade to protect the farmers and their families who settled there. This would encourage others to join them, and they could possibly raise enough corn and wheat to supply both the mining camp and the presidio. Ten or twelve soldiers would accompany the prisoners to prevent them from escaping and protect the men from surprise attack by the Apaches. A corporal would oversee the work, and an officer from the Santa Rita presidio could visit periodically to check on their progress.[4]

Calvo ordered Captain Ponce de León to take the necessary steps to carry out Colonel Arce's plan for the Mimbres settlement. The captain could not possibly obey such an order. Only forty-eight soldiers now manned the Santa Rita presidio, with two or three soldiers deserting each month. Convicts serving time in the penal colony escaped at the same rate as the soldiers left their posts. Both soldiers and civilians were always prepared for Apache raids, even attending mass with weapons in hand. Presidial troops escorted the freight wagons, carried the mail, performed sentry duty, and patrolled the outermost areas of the settlement. To deplete the forces by sending a dozen men to the Mimbres would leave the mining camp even more vulnerable to attack. They never constructed the fortification on the Mimbres.[5]

The price of copper dropped in February of 1837, when the Centralist government ordered all states circulating copper coins to reduce their value by one-half. Copper coins had provided most of the currency in petty dealings since 1814. Each state minted their own coins. Because the value of the copper coins differed in each state, the coins of one state were worthless in another. This caused the merchants a great deal of trouble and annoyance as they accumulated thousands of dollars of worthless coins.

Copper coins of Durango and other states began to depreciate in 1834. At that time Calvo banned the circulation of out-of-state coppers in Chihuahua, thus keeping the value of that state's copper coinage stable. Three years later, when the government ordered all states circulating copper coins to reduce their value, Calvo, as governor of Chihuahua, tried to ignore the order. He told the treasury officials that the public had a high regard for these coins, and no merchant ever refused to accept them. Common people needed this small change to purchase goods and conduct business. The treasury secretary disregarded Governor Calvo's protest, ordering him to obey the law. This was a blow to Chihuahua's fragile economy. Calvo resigned the office of governor of Chihuahua because of the conflict over this issue, although he continued to serve as military commander until his death on February 28, 1838.[6]

Steven Courcier was an astute businessman. He knew that the time was coming when the Santa Rita copper mines would no longer show a profit. He and McKnight had recently acquired a silver mine at Corralitas, a new mining area between Janos and Casas Grandes. McKnight moved some of his Santa Rita workers and their families to Corralitas, leaving Louis Dutton in charge of the copper mines. Although he continued to serve as justice of the peace and mine administrator at Santa Rita, McKnight spent much of his time at the new silver mine.[7]

Meanwhile, some ninety miles south of the copper mines, an encounter between Apaches and Americans had grave consequences for Santa Rita del Cobre and the entire frontier. John Johnson, an American living in Sonora, led a party of seventeen Americans into Apachería. According to Johnson's report of the incident, he had permission from the Sonoran government to pursue the Apaches following their raid on Sonora. The permit entitled the party to keep half the goods and animals they recovered from the Indians.

After searching for several days, Johnson picked up a cattle trail which led to a large Apache camp at a spring near the southern end of the Animas Mountains. Juan José Compá, his brother Juan Diego, and chieftains Marcello and Vivora, with eighty armed warriors and around two hundred women and children, were in the encampment. According to Johnson, he told Juan José they were on their way to Santa Rita. He offered to trade flour, sugar and gunpowder to the Indians for a guide to get them to the copper mines safely. Juan José agreed to this arrangement and even allowed Johnson to ransom a captive, who told Johnson that the Apaches planned to accept the trade goods, then lead the party into an ambush. Johnson and his men decided they had no choice but to attack first if they wanted to escape with their lives. These Indians were responsible for the recent raid on Sonora, so Johnson felt entirely justified in deceiving them.

Johnson's party had a small swivel gun packed on a mule. On the morning of April 22, when the Indians gathered to receive the promised flour and sugar, Johnson's party opened fire with the concealed swivel gun and their American rifles. A two-hour battle left Juan José and Juan Diego Compá, Marcello, thirteen braves and a few women dead, with many more Indians wounded. The remaining Indians fled into the mountains and regrouped. Johnson and his men headed for Janos pursued by the Apaches, who attacked them along the way. The Americans killed seven more warriors. However, Johnson's account does not say how many Americans the Indians killed or wounded in the battle. The Chihuahua newspaper published Johnson's description of the battle and he became something of a hero. Commandant General Calvo recommended that he and his men each receive one hundred pesos for their services.[8]

Told and retold on the frontier, John Johnson's encounter with the Apaches grew into a legend of epic proportions. The battle became a massacre, often said to have occurred at Santa Rita del Cobre, with several hundred Apaches killed. Various accounts said the "massacre" so enraged the Apaches that they vowed to kill every white man on the frontier, starting at the copper mines.[9]

Troop strength at the Santa Rita presidio was forty-two soldiers in the summer of 1837. Five of these men were in the guard house for various infractions of the rules. Only seven convicts were added to the penal colony over the next year. Getting food to the camp continued to be a problem, but somehow the people survived. Although military strength was down, both soldiers and civilians were armed and alert to possible raids. The cannon, mounted behind the parapet of the presidio, could inflict sufficient damage to discourage a band of wrathful Apaches.[10]

The Apaches were masters at storing up crimes and injustices committed against them. They only waited for the right opportunity to seek revenge. Pisago Cabezón bided his time. The key to Santa Rita's survival was the food supply. Courcier sent a herd of cattle and mules and ten wagons loaded with food and supplies to the beleaguered mining camp in late March 1838. A military escort and several armed Americans accompanied the wagon train. Three hundred or more Apache warriors waited to ambush the convoy as it neared Carrizalillo Springs, midway between Janos and Santa Rita. The men guarding the wagon train had no chance against a horde of mounted Apaches armed with rifles and lances. Mexicans and Americans took cover some distance away. Indians ran the cattle and mules off and took over the water hole. The opponents exchanged sporadic gunfire into the next day. Pisago Cabezón was ready to discuss a truce when one of his group reminded him of the treacherous attack on members of his family at Santa Rita. Some Indians wanted to kill everyone in the wagon party. After much discussion, Pisago ordered the Mexicans to abandon everything and return to Janos. He not only granted the men safe passage to Janos, but also allowed them to keep twenty-two horses for their journey.[11]

Apaches raided the Santa Rita camp soon after the attack on the supply train. The Indians killed or wounded several inhabitants and stole horses, mules and a flock of sheep. Five Apaches stopped a Santa Rita wagon train loaded with copper bound for Chihuahua a few weeks later. They were disappointed that the cargo was metal instead of food. The Indians allowed the party to continue their journey, promising to catch them on the return trip when they were loaded with flour, sugar and other foodstuffs.[12]

Meanwhile, the plight of the inhabitants of Santa Rita del Cobre remained grim. Day after day the anxious inhabitants of the mining community awaited the arrival of wagons or mules with supplies. No one came. Captain Ponce de León implored the commander at Janos to send supplies and food to save the lives of the soldiers, miners and families, who faced a "horrible catastrophe." He advised the lieutenant to use all available troops to guard the wagons. "Due to the violence expected en route," he said, "use the strongest animals for the wagons and your troops, even if you have to obtain some from the local authorities." Seventy armed men escorted the next supply train to Santa Rita when it left Galeana on May 6. The wagon train arrived at the mines safely, although they encountered a few Apaches along the way. These were the last supplies to reach Santa Rita. The Apaches had almost won the battle for Apachería.[13]

The final abandonment of Santa Rita was inevitable. Simón Elías González, governor of Chihuahua and soon to be Commandant General, addressed the citizens of that state by proclamation on May 18, 1838. He referred to the damage done by the Indians and the failure of the citizens and government to subdue the savages and establish a long-lasting peace. He made clear the importance of the Santa Rita del Cobre establishment when he said that the imminent abandonment of that mining camp, which served as a barrier to Apache invasions, would leave the settlements across the frontier increasingly exposed to Indian depredations.[15]

The last correspondence from the mining camp was a letter written by Sergeant Todocio Madrid to the commander at Janos, on May 20, 1838. The military commander had removed the Janos Company and their families from Santa Rita, and only a small detachment of soldiers remained. Madrid stated that he sent a discharged soldier to Janos because no jobs were open at Santa Rita to earn enough money for food. Robert McKnight had decreed that only soldiers, convicts and the people directly involved in the mining business would remain in the camp. Madrid stated that no money remained in the military account to ration the troops and prisoners. He could not even keep the daily report because he did not have paper to write on.[16] McKnight was shifting operations to Corralitos, and he had already moved many of his workers and their families to the new place.

In June of 1838, the Apaches added the final blow to the Santa Rita mining operation with another attack on a supply train carrying food and other necessities. The Indians stopped the wagons in a canyon leading to the mines. They took the mules and horses, appropriated what they wanted from the contents of the wagons, and destroyed the rest. The Apaches sent word to the inhabitants at the copper mines that they would allow no more supplies to reach Santa Rita. They threatened to destroy the inhabitants at every opportunity if they did not leave. Hostile Apaches surrounded the camp. Supply trains could not get in, and the copper could not get out. Miners and soldiers were eager to abandon the place. Fleeing miners left a large quantity of charcoal and several hundred dollars worth of ore ready for the furnaces. The exodus was hasty, but no evidence suggests that death and destruction were involved in the withdrawal.[17]

McKnight took his people to Corralitos and put them to work at the new silver mine. Lewis Dutton, James Buchanan, James Kirker and several other Americans from Santa Rita moved to Corralitas with McKnight. Several years later, Buchanan declared that McKnight and Courcier cleared five hundred thousand dollars the last year they worked the copper mines, although they lost an entire mule train loaded with copper to the Apaches that year. Buchanan also stated that gold found in the copper paid for all the expenses of extraction and transportation.[18]

In the early 1840s, the Apaches tried to ensure that the mines of Santa Rita del Cobre would never operate again. When Chihuahua proposed a peace treaty, the Apaches said they would agree to peace only if the Mexicans met certain conditions. They asked that the area of the copper mines be given to the Apaches as their land, and the Mexicans never again settle there. They also stipulated that Robert McKnight and James Kirker would never return to the area. They demanded possession of several of the best water holes and permission to freely sell horses, mules and plunder. These terms were unacceptable to the Mexican government, but it made no difference because the Apaches had won the battle. The copper mines would not be worked again until 1857. By this time, Apachería and its inhabitants were the responsibility of the United States.[19]

McKnight and Courcier obtained a lease on a property known as Veta Grande in the Escondida mountains south of Corralitos in 1840. McKnight located the Barranca Colorado mine, which became well known as a silver producer, and their partnership continued to prosper. Robert McKnight died at the Corralitos hacienda in 1846. Steven Courcier died a few years later, leaving a large fortune and a fine hacienda southeast of Chihuahua to his only son.[20]

Dr. A. Wislizenus heard of the Santa Rita mines as he toured northern Mexico with Colonel Doniphan's expedition in 1846. According to Wislizenus, Courcier was generally believed to have cleared about half a million dollars on the copper mines in seven years. He added: "These copper mines are claimed by the state of Chihuahua, as belonging to its territory; but as not even the latitude of the city of Chihuahua has been well determined by the Mexicans, more exact astronomical observations may perhaps prove that they fall within the territory of New Mexico. This question may become of importance, because this whole range of mountains is intersected with veins of copper and placers of gold."[21]

For more than thirty years, the remarkable mining camp of Santa Rita del Cobre produced abundant amounts of copper under the most difficult conditions. The presence of a commercial venture in the midst of hostile Apache country was unusual enough; the fact that it was referred to as the "watchtower of the western frontier" indicates that the copper mining camp also played an important role in the war against the Apaches. The story of the beginning years of Santa Rita del Cobre is a tribute to the courage and determination of the Spaniards, Mexicans and Americans who held what they had gained despite formidable odds, and to the Apaches who were equally determined to win the struggle for survival.[22]

Epilogue

A lot has happened at Santa Rita since the end of our story in 1838. Kit Carson, the only one of our characters to come back to the copper mine, served as a guide in 1846 for General Stephan Watts Kearney and his "Army of the West." They marched through the old copper camp on a sunny Sunday afternoon, October 18, on their way to California to fight in the war with Mexico. He found the copper mines almost exactly the way they were left eight years earlier.

Had any one of the early-day inhabitants had a chance to visit Santa Rita 150 years later, they would have been astounded beyond belief at developments. Where once lay the lively village of El Cobre, was now a giant open-pit mine 1.5 miles across and 1,500 feet deep. Alongside the new "space" of the pit have risen massive, mountain-sized waste dumps that now produce precipitated copper. Santa Rita's past visitors, however, would still recognize the Kneeling Nun who vigilantly guards the reformulated horizon.

After intermittent interludes of activity in the half century after the 1838 abandonment, miners produced copper at this ancient site almost continuously over the next 170 years. There was a three-year period of mining just before the Civil War and then another four-year stint of operations immediately after the national congflagration. Prior to the rise of modern mining in the twentieth century, several investors made a go of it in the frontier period. A Texas-Mexico partnership initiated the post-Mexican era, restarting the abandoned works in the decade after the American Civil War. The Americans dominated the scene for the remainder of its history after a controversial fight for control pitted members of the famed Santa Fe Ring against the Spanish claimants to the mines (the Elguea heirs). Joseph S. Wilson, a United States land commissioner, rightly granted the copper prospects to the Spanish legatees in 1870 via the provisions of the

Treaty of Guadalupe Hidalgo and territorial mining laws that guaranteed the land claims of former Mexican citizens. Three years later Colorado investor Mathew "Matt" Hayes bought out the heirs, introducing gilded age industrialism to southwestern New Mexico Territory.

The Santa Rita open pit mine. All evidences of the mining camp of Santa Rita del Cobre and the fort that guarded it are gone. The Kneeling Nun, although much reduced by erosion, still watches over the scene. Courtesy of New Mexico State University, Rio Grande Collection.

Hayes' efforts stalled by 1881 when another financier, eastern magnate J. Parker Whitney, acquired the much-touted copper deposits. Poor management and isolated conditions stifled this Bostonian's efforts even though he formed two companies—the Santa Rita Copper & Iron and the Carrasco Copper companies—in hopes of making financial gains. A disappointed Whitney sold out in 1899 to the recently formed Santa Rita Mining Company. This company also failed to make the copper mines pay big, with leasers dominating the mining terrain until everything changed in 1909.

Despite these failed attempts at developing Santa Rita's "red gold," the arrival of the railroad in Grant County in 1881 and then to Santa Rita in 1898 greatly improved the chances for profits at the New Mexico properties. Mining engineer John M. Sully noticed the potentially profitable copper deposits in 1906 when he evaluated the property for the General Electric Company. Though GE ignored Sully's optimistic report of a promising future for the mines, Boston and New York investors took notice. Despite some obstacles, Sully and others convinced Hayden, Stone, & Company of New York City to finance a new venture, the Chino Copper Company.

With the blessings of Daniel C. Jackling, mastermind of the Kennecott Copper Corporation's open-pit operation at Bingham Canyon in Utah, Chino soon began investing in economies of scale technologies to begin open-pit mining at Santa Rita. By 1910 the new corporation brought in large steam shovels to begin digging the pit, steam locomotives to pull the railroad cars of waste and ore, and steam and gasoline drills to punch holes for the explosives. This industrial configuration, though on a much larger scale, is basically the same strategy employed today to mine, mill, and smelt minute flakes of copper imbedded in the earth's top layer. What once took days and even weeks to do now takes only hours (and in some cases minutes) to complete. At the beginning of pit operations, a steam shovel could carry about three tons of material, for example, and today the giant dippers scoop ninety tons of ore. In the early days, multiple rail locomotives pulled many ore cars to transport the rich material that now takes only one giant 290-ton diesel dump truck, usually overloaded with 300 tons per load.

With the massive amount of material that is moved in the copper pit today, there is enough copper produced daily to exceed the entire year's production of a hundred years ago. A good example would be to compare the Santa Rita Mining Company's peak 1903 production of six million pounds of copper to the present owner's 2007 output of three hundred eighty million pounds. Copper prices have also risen alongside production from a low of five cents a pound in 1933 to four dollars in 2007, making profitability a long-term reality at Santa Rita.

Though the fabled company town of Santa Rita and its predecessor El Cobre no longer exist, the future still holds promise at the Chino Mine. The current owner of the property, Freeport-McMoRan Copper & Gold, predicts at least ten more years of production from the longest continuously mined copper deposits in North America. The Spanish, Mexican, and frontier Americans who made Santa Rita a success two hundred years ago would be astonished at the giant chasm that once was Santa Rita and the untold profits made from the native copper outcroppings. The immense open-pit will remain for centuries as a reminder of the incredible achievements of past generations.

Notes

Preface

1. Calvo, Chihuahua, to 2nd Division, November 28, 1835. AHJA f37 s1 p4.
2. Clemons, Christiansen, James (1980) 26.

Chapter 1

1. Griffen (1988) 2; Sweeney (1998) 3-9; Cortés (1989) 339-340.
2. Matson, Schroeder, ed. (1957) 32: 345-346.
3. Schroeder (1974) IV: 1-2.
4. Bernardo de Miera y Pacheco, a native of Spain, was a man of many talents. During his long career, he served as soldier, engineer, astronomer, armorer, sculptor and painter, merchant, rancher and miner. However, he is best remembered for the detailed maps he made of New Mexico and adjacent areas. Miera served as a captain in the cavalry of Cantabria before emigrating to the New World. He lived in the city of Chihuahua for a time, moving to El Paso in 1743. There he operated a mercantile business and was a captain in the militia. In 1747 Miera served as cartographer on a military campaign from El Paso through Apachería to the land of the Hopi. Two years later he plotted the course of the Rio del Norte from El Paso to La Junta. In the 1750s, Miera became alcalde mayor of Pecos and Galisteo and participated in several offensive expeditions against the Comanches. In 1756 the governor of New Mexico commissioned Miera to produce a map of the territory. Incorporating all the information gathered on his travels, Miera completed his elaborately decorated map in early 1758. He remained in Santa Fe, serving for a time as alcalde mayor and producing a plat of the capital. In 1776 he accompanied the Escalante-Domínguez expedition on its search for a northern route to California. Three years later he produced another map of New Mexico for the governor. He died in Santa Fe in 1785. Briggs (1976) 31, 182-83, 187; Kessell (1979) 385-86, 389, 507-08.
5. The watering place known as Paso de Todos los Santos occupied a pleasant valley where the headwaters of the Gila river broke out of the Mogollón Mountains. The spot became well known as a rendezvous point for Spanish troops pursuing Indians into Apachería. As early as 1744, Sonoran governor Vildosola suggested that a one-hundred-man presidio be established on the Gila river which would exert some control over the marauding Apaches.

Paso de Todos los Santos was one of the sites recommended, although no Spanish settlement was ever made in this area. Navarro Garcia (1964):89-90; Bolton (1950) 247.

6. Navarro García (1979) 207; Cortes(1989) 34-46.
7. Thomas (1941) 30-31; Thomas (1932) 5-6.
8. Griffen (1988a) 11-13; Spicer (1962) 239.
9. Park (1962) 4: 328; Thomas (1941) 24-27.
10. Park (1962) 340-341.
11. Griffen (1988a) 44.
12. Cortes (1989) 6-7; Park (1962) 4: 340-341.
13. Griffen (1988a) 79.
14. Ibid. 76.

Chapter 2

1. Griffen (1988a) 38.
2. Almada (1968) 89; Almada to Lundwall, April 16, 1968.
3. As part of the peace program, the Commandant General encouraged his presidial officers to make friends with the Apache leaders, learn their language and customs, and treat them with courtesy and patience. Cortès (1989) 7.
4. Carrasco to Tovar, Real del Cobre, November 30 1804, AAC491 r154;
Carrasco to Deputatión de Minería, Cobre, January 3, 1805, AAC491 r154.
5. La Bufa literally translates as a comic female singer. It was also a name often used in northern Mexico for a peculiar rocky formation crowning a tall hill. See Appendix for "The Legend of the Kneeling Nun."
6. From later descriptions of the site, geological reports, and a bit of conjecture, we can visualize Santa Rita del Cobre as Carrasco first saw it. Lindgren, Graton, Gordon (1959) 199; Baltosser (1968) 21-23.
7. Barrett (1987) 2, 43-62.
8. Bartlett (1854) 179, 230.
9. Carrasco to Tovar, Cobre, November 30, 1804, AAC491 r154.
10. Haring (1947) 264, 272; Brading (1971) 150.
11. In 1776, in an effort to find a solution to the hostile Indians who were devastating New Spain's northern frontier, Spain created the *Provincias Internas* or internal provinces. Composed of the northern provinces and the territory of New Mexico, the Provincias Internas were separated from control of the Viceroy and governed by a military official known as the Comandante General. Jenkins, Schroeder (1974) 29.
12. Carrasco to Tovar, Cobre, November 30, 1804, AAC491 r154.
13. Nava to Valle, Chihuahua, November 5, 1800, AHJ498 r14.
14. Libro de Registro de denuncios mineros presentados ante la Diputación Territorial de

Minería, Estado de Chihuahua, pagina 17 frente, June 30, 1801. AAC491 r151. Each large mining center had a territorial Deputación de Minería composed of a judge and two deputies elected from qualified mine owners and administrators. Their duties were to enforce the mining ordinances, settle disputes, inspect mines, and promote the interests of the miners. Halleck (1859) 201-206.

15. Barrett (1987) 24.
16. Libro de Registro de denuncios mineros, Chihuahua, June 30, 1801, August 8, 1803, January 24, 1805, July 30, 1806.
17. Baltosser (1968) 21-23.
18. Receipt for eighty-four arrobas of copper, issued by Pedro Ramos de Verea, Chihuahua, July 22, 1801, AHJ498 r14.
19. The Bavispe presidial company of loyal Ópata Indians was organized in 1799. They served as scouts, defended Indian villages and pueblos, and accompanied Spanish troops on campaigns against the Apaches. Although they presented a fierce appearance, they were noted for their good disposition, discipline, sturdiness and good marksmanship with muskets or bows and arrows. They frequently provided protection for the miners at El Cobre and served as escorts for the mule trains during the Spanish Colonial and the Mexican periods. Moorhead (1965) 92; Cortes (1989) 26-27.
20. Carrasco to Tovar, November 30, 1804, C154 r5.
21. Carrasco to Deputación de Minería, January 3, 1805, AAC491 r154.
22. The strong support of Commandant General Salcedo was vital to the success of the Sierra del Cobre mines in these formative years. Nemecio Salcedo y Salcedo was a native of Spain, born in Vizcaya in 1750. He pursued a military career, coming to New Spain as a Colonel commanding the Royal Infantry Regiment. The Viceroy considered him an "officer of much talent and learning, with untiring devotion to duty." Salcedo assumed the office of Commandant General of the Provincias Internas in November of 1802. He held this office until July of 1813 when he turned the command over to Bernardo Bonavía and returned to Spain. Navarro García (1965) 3; Archer (1977) 259.
23. Libro de Registro de Denuncios Mineros, Chihuahua, August 3, 1803, p.19 frente, AAC491 r151.
24. Salcedo to Janos, March 18, 1804, AHJ498 r15.
25. Bowden (1971) 11.
26. "San Vicente de Paul 1874-1974," an address given by Catholic historian Rev. F. Stanley during the 1974 centennial celebration of San Vincent de Paul Catholic Church in Silver City, New Mexico. Father Stanley gave only vague references to the source of this story, and no documentation has been found to confirm it. There may be some basis of truth, however, as tradition says that there were Mexican people at the ciénega long before there was a mining camp there. Evidence of early mining and smelting, predating the founding of Silver City, was later found in the area. *The Borderer*, Sept. 21, 1872; *Silver City Enterprise*,

Sept. 19, 1902; *Silver City Independent*, March 3, 1908. One of the earliest maps of Silver City, drawn in 1874 by A. Z. Higgins of Santa Fe, shows an old fortification on the hill above the valley.

27. Tovar, Janos to Salcedo, July 19, 1803, AHJA f17 s3 p1.

28. Carrasco to Deputation de Mineria, Jan.3, 1805,AAC491 r154.

29. Francisco Manuel de Elguea, Chihuahua, to Deputation de Mineria, January10, 1805, AAC154, 8a&b; Deputation de Mineria, March 23, 1805, AAC154 10d; Jose Manuel Carrasco, El Cobre, to Deputation de Mineria, November 16, 1810, AAC154; Antonio Cordero, Durango to Commandant of Janos, March 5, 1819, AHJ498 r12; Janos Commandant, Janos, to Commandant General, March 29, 1919, AHJ498, r12.

Chapter 3

1. Almada (1968) 184.

2. Moorhead (1958) 53 n48; Pablo Guerra, Chihuahua, to Ayuntamiento of Chihuahua, May 17, 1826, AAC491 r176.

3. Baxter (1982) 8, 86; Maynes, Janos to Commandant General, January 16, 1817, AHJ498 r19.

4. Nemecio Salcedo to Chihuahua commission, Chihuahua, October 24, 1805. AAC491 r145.

5. Humbolt (1808) vlll 260-61; Obeso, Cobre, to Salcedo, September 30, 1804, AHJ498 r15.

6. 1805 map of the Provincias Internas, ordered by Commandant General Salcedo produced by Lt. Juan Pedro Walker of the Presidial Company of Janos. Barker Texas History Center, University of Texas at Austin. Rickard (1923) v116 #18, 758.

7. Salcedo, Chihuahua, to Ochoa, April 6, 1803. AHJ498 r15.

8. West (1949) 19-25; Bakewell (1971).131-32; Humbolt (1808) IV: 238.

9. Kemp (1972) 6-10; A large section of timbering with many of the rawhide bindings still in place, was found in an old shaft at Santa Rita in later years. Sully "Chino Copper Company," *Resources and Industries of the Sunshine State*, nd.

10. Hendricks (1999)152-155.

11. Salcedo, Chihuahua, to Soler, May 9, 1804, AGI Guadalajara, 296 #59.

12. Salcedo, Chihuahua, to Janos commander, February 6, 1805, AHJA f18 s1 p3.

13. Salcedo, Chihuahua, to Janos commander, November 26, 1803; February 3, 1804, and April 7, 1804, AHJ498, r15.

14. Salcedo, Chihuahua to Soler, May 5, 1805, AGI Guadalajara, 296 #87; November 12, 1805 #95; March 4, 1806 #106.

15. Tovar, Santísima Trinidad del Oro, to Salcedo, November 30, 1804. AHJ498 r15.

16. Libro de Registro de Denuncios Mineros, p21 frente, Chihuahua, July 5, 1801, AAC491 r151; Thurston, Attwater, ed. (1962) II: 369-70.

17. Tovar, Real de Santísima Trinidad, to Deputación de Minería, November 30, 1804, AAC491 r154; Salcedo to Soler, December 4, 1804, A.G.I. 296 #72.

18. Tovar to Salcedo, December 4, 1804, AHJ498 r17.

19. Salcedo, Chihuahua, to Soler, #87, May 7, 1805; #95, November 12, 1805; #106, March 6, 1806, AGI, Guadalajara, 296.

20. Libro de Registro de Denuncios Mineros, July 30, 1806, p.23, AAC491 r151.

21. Salcedo to Solar, #108, June 1806; #110, September 9, 1806, AGI, Guadalajara, 296.

22. Certified copy of death certificate of Francisco Manuel de Elguea, courtesy of Francisco Almada, April 16, 1968.

Chapter 4

1. Carrasco to Tovar, November 30, 1804, AAC154 4e.

2. Weber (1982) 223-224.

3. Janos to Salcedo, #210, January 16, 1805. AHJ(A)f17 s3. Tradition says that Elguea built a large triangular adobe fort at the mines; this is not true. There is reference, however to "the old torreon near the principal mine." The presidio was not erected until 1835 (see chapter XII for details). Daily Report, June 30, 1836, AHJ498 r27.

4. Tovar, Janos, to Salcedo, December 10, 1804, AHJA f17 s3.

5. Tovar, Real de la Santisima Trinidad del Oro, to Deputation de Minería, November 30,1804, AAC154/4a .

6. Pachitejú was the leader of a band of Mimbres Apaches in the 1770s. He and his father, Natanijú, and his grandfather, Chafalote, lived in the Sierra de los Mimbres and the Sierra del Cobre. Pachitejú and his band probably had a rancheria near the springs, therefore the Spaniards named the area after the Apache chief. Kinnaird (1946) 78; Thomas (1932) 15-16.

7. Tovar, Janos, to Salcedo, Dec. 10, 1804, AHJ498 r15 #204; AHJA f17 s3.

8. Tovar, Janos, to Salcedo, January 16, 1805, AHJA f17s3, #210.

9. Salcedo, Chihuahua to Tovar, February 3, 1804, April 7, 1804 AHJ498, r15.

10. Upper Camp Cobre, December 3, 1804, AHJ498 r15; report from Janos giving location of the garrison, December 10, 1804, 20 soldiers on detached service at El Cobre; Salcedo, Chihuahua, to Janos commander, October 11, 1804, AHJ498 r15 and November 6, 1805, AHJA f18 s1 p33.

11. Salcedo, Chihuahua, to Tovar, Aug. 10, 1804, AHJ498 r15. A comisario de justicia was similar to a frontier justice of the peace. Unfortunately none of these monthly reports have been found in the archives. The records of the Commandant General have been missing for many years; the state archives of Chihuahua were distroyed by fire in 1941. Almada (1938). Some of this information is in correspondence between the Commandant General, the commander at Janos and the comisario de justicia at El Cobre.

12. Salcedo, Chihuahua, to Tovar, Sept. 8, 1804, AHJ498 r15.

13. The mining district was first identified as Santa Rita del Cobre in September of 1805, in the

criiminal case against Julio Carrasco and other accomplices in the theft and disposition of 15 sheep. The name Santa Rita del Cobre was often used in court documents and mining registrations after this date.

14. Cobre, September 1-14, 1805, AHJA f18 s1 p83.
15. Parrish register of Church of Nuestra Señora de Guadalupe, Paso del Norte, August 15, 1806.ACJ513 r6; Teran to Janos, Real del Cobre, July 22, 1807, AHJ489 r15.
16. Mateo Sánchez Álvarez, Chihuahua, to José Miguel Yrigoyen, September 1, 1812; Yrigoyen, Durango, September 19, 1812, AHJ498 r2. Sometime after open pit mining began at Santa Rita in 1910, around fifty skeletons were found on the northern extremity of the Hearst pit. They were lying within a small area and at a depth of about six feet under the natural surface and some fifteen feet below the top of an old dump. No record could be found of the existence of such a burying ground. The workers who attended the shovels were very superstitious about this area and nervous about working there. Sully (1933).
17. Salcedo, Chihuahua to Sosaya, February 23, 1807, AHJA f18 s1 p33.
18. Salcedo, Chihuahua, to Janos commander, July 1, 1807, AHJA f18 s3 p133.
19. Salcedo, Chihuahua, to Janos, February 6, 1805, AHJA f18 s1 #1.
20. Janos commander to Salcedo, June 7, 1808, AHJ498 r15; Teran, Cobre, to Janos commander, April 27, 1809, AHJA f19 s2.

Chapter 5

1. To claim a mine, the discoverer registered it with the Diputatión de Minería. He had ninety days in which to sink a pit or open a trench on the vein to a depth of twenty-seven feet. Halleck (1859)
2. Salcedo, Chihuahua, to Soler, AGI May 5, 1805, 296 #87; Salcedo, Chihuahua, to Solar, AGI296, March 4, 1806 #106; June 1806 #108; September 9,1806 #110.
3. Jackson ed.(1966) 1:409.
4. Ibid., 2:48.
5. Salcedo, Chihuahua, to Solar, May 1805-April 1806, AGI296, numbers 95, 106, 108.
6. Rickard (1923) 754.
7. Barrett (1987) 64 and 80.
8. Ibid., 88.
9. Rickard (1924) 13.
10. Agricola (1950) 355-90; West (1949) 27-28.
11. Barrett (1987) 89; Rickard (1924) 13.
12. West (1949) 42.
13. Salcedo, Chihuahua, to Francisco de Saavedra, #8, June 6, 1809. AGI Guadalajara, 429.
14. Viceroy Pedro Garibay, February 7, 1809; Viceroy Francisco Xavier de Lizana y Beaumont, August 28, 1809, ANM r22, frame 453 & 464.

15. Ibid.

16. Salcedo, Chihuahua, to Francisco de Saavedra, #8 June 6, 1809, AGI Guadalajara, 429.

17. Salcedo, Chihuahua, to Janos, June 8, 1809, AHJA f19 s2 p91; Teran to Janos, Cobre, Aug. 29, 1809.

18. Barrett (1987) 2-4, 51-53.

19. Census of Cuidad Chihuahua, 1822, AAC491 r141; census of the Hacienda del Torreón, 1826, AAC491 r152.

20. Fr. Agustine Paltan, chaplain of Janos presidio, obtained permission to erect a chapel, Nuestra Senora del Pilar, at the west end of Janos. It was completed by 1808. The bells must have been ordered for this chapel, as the dates coincide. Griffen (1988a) 101; Teran, Cobre, to commander at Janos, June 16, 1809, AHJ) f19 s2 p223.

21. Howell's letter published in The Ashtabula Sentinel, June 1858, copy courtesy of Professor Thomas Wortham, University of California at Los Angeles, who brought this story to my attention.

22. Meyer, Sherman (1979) 281.

23. Teran, Cobre, to Janos, July 15, 1809. AHJA f19 s2 p205.

Chapter 6

1. Teran, Cobre, March 29, 1810, AHJ498 r1; No background information was found for José Baca, but his ability to do the job well over a long period of time is proved by the success of the Santa Rita operation during the rebellion.

2. Salcedo, Chihuahua to Janos, March 21, 1813, AHJ498 r2.

3. Francisco Chavez vs Elguea estate, December 6, 1812-January 14, 1814, AAC491 r133.

4. The San José copper mine was rediscovered and profitably worked in the early 1860s. It was again abandoned during the Civil War and relocated in 1869 by James D. Fresh, superintendent of the Santa Rita mines. Fresh and his partner, John Magruder, built a smelting works in 1873 and took out a considerable amount of ore. The mine passed through many hands during the next few years. San José was a booming mining camp during the 1880s, then failed as the ore became low grade. In 1928 the property became part of the Ground Hog mine owned by the American Smelting and Refining Company (ASARCO).

5. Salcedo, Chihuahua, to Janos, March 22, 1810.

6. Baca, Cobre, to Janos, May 23, 1810. AHJ498 r1. Baptism, marriage and burial records performed in Mineral de Santa Rita del Cobre by Father Rafal Echeberría. AHJ r27.

7. Salcedo, Chihuahua, to Baca, August 22, 1810 AHJ489 r11.

8. Almada (1968) 37-38.

9. This division of territory was used in treaty negotiations through the Spanish Colonial period and many years later as the Mexican government struggled to deal with the Apache. Calvo, Encinillas, July 28, 1832, AHJ498 r5.

10. Salcedo, Chihuahua, to Janos, Oct. 18, 1810, AHJ498 r11.

11. Salcedo, Chihuahua, to Janos commander, January 12, 1812, AHJA f20 s3 p9; Janos to Baca, March 1812, AHJA, f20 s3 p159.

12. José Baca, Real del Cobre, to Captain José Ronquillo, April 25, 1812, AHJ498 r2.

13. Sweeney (1998) 403 n7 and n8; Griffen (1988a) 127.

14. Bonavía, Durango to Janos, August 30, 1814, AHJ498 r18; Baca, Cobre, to Bonovía, circa October of 1814, AHJ498 r2.

15. Griffen (1988a) 94; Sweeney (1998) 37.

16. Sweeney (1998) 27-28.; Baca, Cobre, nd, AHJ498 r2. 192.

17. Salcedo reports to AGI 1809-1813; Guias issued for transporting ore 1813-1820.

Chapter 7

1. Almada (1968) 243. Copy of marriage certificate courtesy of Francisco Almada, April 16, 1968.

2. Tesorería General del Estado. Archivo de la extinguida Administración General de Rentas. Certified copy, Francisco Almada to H. Lundwall, April 16, 1968. Guías quoted from 1813 to1833 are from this list provided by Francisco Almada.

3. Caballero (1974) 55-56.

4. Green (1987) 216.

5. Ibid., 112-130; Staples (2003)151-154; Caballero (1974) 55-57; Green (1987) 128, 216.

6. Meek (1948) 71-73.

7. Barrett (1987) viii.

8. Salcedo report, August 11, 1812 and February 9, 1813, AGI 429 #94, #134.

9. Tesorería General del Estado. Archivo de la Extinguida Administración General de Rentas. Copy courtesy of Francisco Almada, April 16, 1968.

10. n/s, April 2, 1819. AHJA f24 s2; William M. Pierson, U.S.Vice-Consul, Paso del Norte, Mexico, December 6, 1873.

11. Meek (1948) 73.

12. Ore shipping permits from Santa Rita del Cobre, February through August 1818; n/s April 2, 1819; Guía issued to Juan Álvarez, January 4, 1820; Guía issued to Geronimo Maceyra, September 4, 1820.

13. Guerra to Chihuahua City council, May 24, 1822.

14. Viscarra, Janos to Conde, June 30, 1823, AHJ498 r17.

15. Commander of First Division, El Cobre to Commander of Army, April 30, 1824. AHJ r16; Sweeney (1998) 39-40.

16. Guía issued to Pablo Guerra, January 14, 1826, covering 128 ingots of copper weighing 790 arrobas (20,000 lbs.) in bars of 156 pounds each, consigned to Manuel Merion, Mexico City. Copy courtesy of Francisco Almada, April 16, 1968; Pattie (1962) 66.

17. Caballero (1974) 97-111.

18. Elguea's son, Francisco, is mentioned frequently in reports of the Santa Fe trade in the 1840's. In 1846 Francisco and several other Chihuahua merchants arrived at Independence, Missouri, with $350,000 in specie (money in coin) and 1,000 mules. He returned to Mexico loaded with goods, only to be caught by the advance of United States troops on Mexico. Elguea and several other merchants were forced to follow Colonel Doniphan's army to Chihuahua. Gibson (1935) 325-328; Barry (1972) 527, 580, 639; Walker (1966) 136. Mariano Elguea died in Chihuahua in 1842, leaving his share of the Elguea estate to his natural daughter, who was his only heir. Records of the Surveyor General, New Mexico Land Grants, File #107, New Mexico State Records and Archives. Mariano Elguea to Ayuntamiento de Chihuahua, June 13, 1827 AAC491 r153.

19. Willard (1962) 249; Batman (1986) 7; Barry (1972) 119.

20. 1826, folio 29, #482, Francisco Almada, April 16, 1968.

Chapter 8

1. Pattie (1962).

2. Morgan (1964) establishes Pattie's arrival in New Mexico at 1825; Batman (1986) unravels a number of inconsistencies in Pattie's story.

3. For background on Pattie family see: Pattie (1962) 5-9; Batman (1986) 22-47.

4. Weber (1971) 52-53, 98-99.

5. Caroll, Haggard (1942)105; Resendéz (2004) 118-119.

6. Pattie (1962) 47-66. The archives of Janos do not mention the presence of any foreigners at the copper mines; therefore, we must rely on the narrative of James Pattie to piece together the story of Santa Rita del Cobre for this period.

7. Ibid., 67.

8. Ibid., 68.

9. Ibid., 70.

10. Ibid., 68-72.

11. Copy of guías courtesy of Francisco Almada, April 16, 1968.

12. Weber (1971) 118.

13., Ibid., 118-119.

14. Pattie (1962) 74.

15. Weber (1971) 137; Warner (1907-08) 183; Foster (1887) 30; Batman (1986) 193.

16. Pattie (1962) 74.

17. Hardy (1977) 462-463.

18. Pattie obtained wine and spirits from El Paso to sell at Santa Rita del Cobre and manufactured goods brought over the Santa Fe Trail from the United States. Hardy confirms the enormous profits possible for mine owners from this mercantile trade. Ibid., 177 -178; Pattie (1962) 113.

19. Ibid., 118-120. August 19, 1827, guía #40 issued in Santa Fe to Santiago Glenn, and Santiago O. Pattie as clerk in charge of the goods. Weber (1967) 32, citing MANM

20. September 22, 1827, guía #44 issued to Sylvester Pattie permitting him to journey to Chihuahua and Sonora as a trader. Weber (1967) 32; Pattie (1962) 121.

Chapter 9

1. Robert McKnight, James Baird, Samuel Chambers, Benjamine Shreve and Michael McDougal provided the trade goods. Thomas Cook, William Mines, Alfred Allen and Peter Baum went along for the adventure. They were accompanied by Charles Myette Cuado as guide and interpreter. Golley (1959) 178-179.

2. Ayudante Inspector to Administrator of Public Works, October 3, 1814, AAC 149, 3; Jones (1988) 182.

3. Strickland (1965) 259-268; Golley (1959) 174, 189.

4. Hardy (1977) 4 77; AAC491 r156; *Diccionario Porrua de Historia, Biografía y Geografía de Mexico,* 541; Timmons (1980) 84:1 9-10.

5. Pablo Guerra was issued a passport to leave the Republic of Mexico as ordered under the law of December 20, 1827. CZ505 r4. In filing a denuncio on abandoned property, the claim must state the location of the mine, its last owner, and the circumstances under which it was deserted, if known. The owner was to be notified and must appear within ten days to retain ownership of the claim. Halleck (1859) Title VI article 8, 225.

6. Letter from U.S.Vice-Consul William M. Pierson to Assistant Secretary of State William Hunter, Paso del Norte, Dec. 6, 1873, in "The Santa Rita Native Copper Mines," 31. This promotional publication was written by, or at the request of, J. P. Whitney, who owned the Santa Rita mines from 1880 to 1899; Escudero (1834) 143.

7. *Archivo General de Notarias Dependiente del Gobierno del Estado de Chihuahua* 68:177 y siguientes. Almada to Lundwall, April 16, 1968.

8. John McKnight made many trips between Missouri and Chihuahua, bringing goods, supplies and equipment over the Santa Fe Trail. He opened a store in Janos in the early 1830's and took advantage of the ore wagons to haul his merchandise. John McKnight was still in the Chihuahua mercantile business in 1846, when he and Francois X. Aubry arrived in St. Louis to purchase goods. They were reported to have brought with them from Mexico between fifty and sixty thousand dollars in specie. Two days after McKnight and Aubry reached Missouri, General Kearny arrived in Santa Fe, where he proclaimed New Mexico a territory of the United States. McKnight did not return to Mexico. Barry (1972) 638.

9. Sam Bean in *Mining Life,* July 5, 1873. Bean was the son-in-law of James Kirker, who was closely connected with McKnight and Santa Rita between 1828-1838.

10. William M Pierson, Paso del Norte, to William Hunter, December 6, 1873; Guia issued to Courcier, March 18, 1834.

11. Louis Dutton was born in New Hampshire between 1802 and 1804. He came to New Mexico in the late summer of 1825, possibly with Robert McKnight when he returned from Missouri. He was an experienced fur trapper and spent his first years in the west in pursuit of beavers. In 1838, when McKnight and Courcier gave up Santa Rita del Cobre and moved their mining operations to Barranca Colorado, near Corralitos, Dutton took up residence there. At the time of the U.S. war with Mexico, Dutton was listed as an American merchant and imprisoned in Chihuahua. After the war he moved to the El Paso area, and by 1860 he was living at San Elizario. During the American Civil War he was accused of being a Southern sympathizer and his property was confiscated. He died at San Elizario in 1875. Strickland, (1965) 9:147-152; Weber (1971) 123.

12. Henry Corlew was born in Kentucky. He came to New Mexico about the same time as Dutton, and may have been one of the merchants in the caravan with McKnight in 1825. He is said to have owned a store in Santa Fe the following year. Before the Civil War he was living in San Elizario and may have been associated with Dutton in the mercantile business. In 1868 Corlew was one of the owners of a gold mine in Pinos Altos, near Santa Rita. The 1870 U.S. census lists him as a merchant, living in Ysleta, Texas. Mills (1962) 13, 176.

13. James Buchanan was born in New York in 1813. After leaving Santa Rita, he became associated with the Chihuahua trade. During the United States-Mexican War, Buchanan was one of twenty American merchants in Chihuahua who were interned until after the war. The United States Census of 1860 listed Buchanan as a resident of Concordia, El Paso County, with a wife and six children. Apparently he died before 1870. Mills (1962) 175.

14. James Kirker, born near Belfast, Ireland, December 2, 1793, came to the United States in 1810. By 1817, he was established in New York City with an interest in a grocery store. He left family (a wife and son) and business behind and moved to St. Louis, where he opened another grocery store and became acquainted with Robert McKnight's family. By 1825 he was trapping for beaver in New Mexico. His name has been linked with Sylvester Pattie at the Santa Rita Copper mines, but other sources say that he came to the copper mines with Robert McKnight in 1828. He made Santa Rita his home for the next eight years, trapping in the winter and working for McKnight the rest of the year. Kirker made a name for himself as an Indian fighter in the late 1830's and early 1840's, when a bounty was offered for Apache scalps. During the war between the United States and Mexico, Kirker joined forces with Colonel Doniphan. His knowledge of northern Mexico and its defenses was of great value in the victory of the American forces. James Kirker died in California in 1853. McGaw (1972); Almada (1968) 302; Smith (1999) 19. Kirker descendents worked in the Santa Rita open pit copper mine for many years. Some of them still reside in Grant County.

15. William M. Pierson, Paso del Norte, to William Hunter, December 6, 1873; Quaif (1966) 8-9.

16. Weber (1971) 221.

17. Salcedo to Janos commander, July 24, 1812, AHJA f20, s3 p47; Weber (1982) 95; McCarty (1997) 39-40.

18. Griffen (1988) 154-158; Sweeney (1991) 21, 23; McCarty (1997) 126-127n2.

19. Governor Madero was well qualified to make this decision. Starting his career in charge of the cashier's office at the Royal Treasury, he served as paymaster and treasurer of the Comisario General. Next he took charge of the general administration of revenue; at the same time he was state treasurer. During his term, he organized the local public treasury, balanced the budget and supplied large sums to the federal treasury in the form of loans so the troops garrisoned in the state could be paid punctually and still had funds in the treasury when he left office. Almada (1968) 318-319; Almada (1950) 34-35.

20. Notice issued July 21, 1831, CZ505 r3; Meek (1948) 71-73.

21. Escudero (1834) 142-145.

Chapter 10

1. Janos to Comandant General, June 30, 1829, AHJ498 r22.

2. Guía issued to Courcier April 17, 1833, copy courtesy of Francisco Almada, April 16, 1968.

3. Almada(1968)15.

4. Notice issued by Governor Madero, Chihuahua, March 1, 1832, CZ505 r3.

5. List of donars, February 20, 1832, AAC491 r163.

6. Ronquillo to commander of Carrizal, Feb. 22, 1832, CZ505 r3.

7. Griffen (1988) 140-141; Sweeney(1998) 48-51.

8. Griffen (1983) 7:2 21-49.

9. Griffen (1988) 146.

10. Jose Saenz, Real del Cobre, to alcalde at Janos, May 10, 1833, AHJ498 r6.

11. Ibid.

12. Ponce de León, San Buenaventure, to Vizcarra, March 24, 1833, AHJ498 r25.

13. Juan José Compá to Mariano Varela, April 25, 1833, AHJ498 r25.

14. Commander of Janos Company to Simón Elías González, June 9, 1834, AHJ498 r26.

15. McCarty (1997) 39-40, 126-127n2.

16. Alejandro Ramírez, Villa del Paso, to Commandant General, October 15, 1834, AHJA f36 s1 p.1.

17. Ibid; Commandant General, Chihuahua, to commander of 2nd Division, January 19, 1835, AHJ498 r26.

Chapter 11

1. In September of 1834, the legislature combined the political and military commands under one person to make the war against the hostile Indians more efficient. Almada (1950) 53.

2. The mines were described as situated in the extreme northeastern corner of Chihuahua, in the Galeana district. Escudero (1834) 142.

3. About the time Santa Rita resumed operations, a Chihuahua newspaper reported that Stephen Courcier's annual income was fifty thousand pesos. This placed him in the top income bracket in Chihuahua City, with only four others of equal status. *El Fanal*, October 28, 1834.

4. Cayetano Justiniani was born in the city of Jalapa, Veracruz in 1800, and entered the military service at the age of twenty-one. He went to Chihuahua in 1828 to serve as secretary to the Commandant General. In 1834 he served as commander of the 2nd Division at Santa Rita del Cobre. The following year he was promoted to Lieutenant Colonel and returned to Chihuahua as secretary to the Commandant General. In 1836 he returned to Santa Rita del Cobre for a short time to try to straighten out some of the problems of the 2nd division. The following year, he was named military commander at El Paso, and in August took a division to New Mexico to put down an uprising against Governor Albino Pérez. In 1838 he became Ayundante Inspector, and served as acting Commandant General several times until he had a serious disagreement with the governor over the use of volunteer forces led by James Kirker to fight the Apaches. He resigned, and retired to his hacienda at Corral de Piedras. In 1840 he again became secretary to the Commandant General, and was made prefect of Hidalgo del Parral at the end of the year. During the war with the United States, Justiniani was Major General of Operations. He served several terms on the State Legislature and held various offices in Chihuahua until his death on June 18, 1863. Almada (1968) 300-301.

5. Inocencio Cisneros, Chihuahua, October 20, 1834, AAC172; Janos Company daily report for November 1834, AHJA f36 s2 p3.

6. Calvo, Chihuahua, to commander, Janos, Nov.7, 1834, AHJ498 r6.

7. Griffen(1988) 157; Calvo, Chihuahua to Justiniani, January 19 & 25, 1835, AHJ r26; Norberto Moreno, Allende, to Chihuahua Ayuntamiento, March 1, 1835, AAC491 r167.

8. Calvo, Chihuahua, to Justiniani, January 19 and 25, 1835, AHJ498 r26; Bartlett, 1854) 1:235; Odie B. Faulk (1969) 23-24.

9. José Hernández, Janos, to Justiniani, June 18, 1835,AHJ498 r28.

10. Calvo, Chihuahua, to Justiniani, May 15, 1835, AHJ498 r28.

11. Justiniani, Cobre, to Calvo, January 6, 1835, AHJ498 r26.

12. Compá, Loma Blanca, to Calvo, January 6, 1835, CZ505 r4; Justiniani, Cobre to Calvo, February 12, 1835, AHJ498 r28.

13. McCarty (1997) 126-127n2.

14. Stevens (1964) 216-17; McCarty (1997)138.

15. List of items sent from Chihuahua for General Juan José Compá, February 25, 1835, AHJ498 r28.

16. Calvo to jefe politico, Cuidad Chihuahua, February 20, 1835, AAC 167 r19; Justiniani to Calvo, June 1, 1835, AHJ498 r6.

17. Simón Elías, Chihuahua to Justiniani, July 9, 1835, AHJ498 r6.

18. Weber (1982) 95-103; Ponce de León, Galeana, to alcalde of Carrizal, February 20, 1835, CZ505 r4.

19. Calvo to commander of western frontier at Cobre, February 27, 1836; Draft of credit, M. Ponce in favor of Esteban Courcier, April 7, 1836, February 29, 1836, AHJA f38 s1 p6.

20. Arce to Justiniani, Aug. 4, 1835, AHJ498 r28; Justiniani, Cobre, to José Maria de Arze, Aug. 14, 1835, AHJ498 r28.

21. Justiniani, Cobre, to Calvo, March 28, 1835, AHJ498 r26; J. Maria Arze, Chihuahua, to Justiniani, March 28, 1835, AHJ498 r26; Calvo to Janos, March 27, 1835, AHJA f37 s1 p14.

22. Tornel, Minister of War, Mexico City, to Commandant General, March 6, 1835, CZ505 r4.

23. Second Field Division, Daily Report, October 31, 1835, AHJ498 r6.

24. Justiniani, Cobre to Calvo, June 4, 1835, AHJ498 r6.

25. Janos to Calvo, December 6, 1834, AHJA f36 s3 p3; Calvo, Chihuahua, to Janos, December 15, 1834, AHJA f36 s1 p3; Janos to Calvo, March 1, 1835, AHJA f36 s3 p17.

26. Calvo, Chihuahua, to 2nd Division, November 28, 1835, AHJA f37 s1 p4.

Chapter 12

1. Espinosa (1967) 82.

2. Matamoros, on the Gulf of Mexico at the mouth of the Rio del Norte (Rio Grande), was a busy port. Chihuahua received nearly half of its manufactured goods, drygoods, and foodstuffs through this port. Many Americans in the trade returned to the United States via Matamoros, where packet boats from New Orleans picked up the wagons and returned them to Missouri. Courcier used this port to receive supplies and ship copper ingots to foreign markets. Major Wetmore to Secretary of War, October 11, 1831, Moorhead (1958) 195n2; Custom house permit to Courcier for merchandise valued at 6475 pesos, April 1, 1837, AHJ498 r27.

3. Jones (1979) 32, 37, 61, 159.

4. Zapata vs. Borsigo, Cobre, April 15, 1836, AHJ498 r27.

5. Janos commander to Justiniani, February 14, 1935, AHJ498 r27.

6. Investigation of Private Albino Bigeria, Cobre, April 15, 1835, AHJ498 r6.

7. Green (1987) 187; Weber (1982) 116-117.

8. Cisneros vs. Hernándes, September 29, 1836, October 1, 1836, AHJ498 r27.

9. Loomis (1958) 265; Calvo, Chihuahua, to Guillermo Dryden, April 8, 1835, AHJ498 r2.

10. Justiniani, Cobre, to Dryden, May 10, 1835, AHJ498 r26; Juan José Bustamente, Chihuahua, to Justiniani, August 10, 1835, AHJ498 r28; Justiniani, Cobre to Calvo, June 15, 1835, AHJ498 r26.

11. Report of Public Works, Mineral del Cobre, April 1836, AHJA f39 s2 p1.

12. Orcacitas, Cobre, to Commander of 2nd Field Div., July 5, 1835, AHJ498 r26.

13. Cisneros, Cobre, to Justiniani, March 26, 1835, AHJ498 r26.
14. Jones (1979) 139-141, 188-189; Alejandro Ramírez, Villa del Paso, to Commandant General, October 15, 1834, AHJA f36 s1 p1.
15. García, Cobre, to Calvo, Dec. 23, 1836, AHJA f38 s1 p19; Cobre, to Janos commander, September 22, 1935.
16. Arce to Justiniani, Aug. 4, 1835, AHJ498 r28; Justiniani, Cobre, to José María de Arce, Aug. 14, 1835, AHJ498 r28.
17. Anastacio de Nava, Chihuahua, to McKnight, June 5, 1835, AHJ498 r6; Kinnaird (1946) 328; Jones (1979) 141, 189.
18. Calvo, Chihuahua to Rey, January 7, 1836, AHJ498 r26.
19. Calvo, Chihuahua, to Ponce de León, January 7, 1836, AHJA f38 s1 p9; Cobre (no signature) to Calvo, February 10, 1836, AHJA f38 s2 p1. During Calvo's administration as governor of Chihuahua, he ceded his governor's salary of 3,500 pesos a year to benefit public education. Several schools were founded through his generosity. Almada (1950); Commandant General Nemecio Salcedo (Commandant General 1802-1813) was also a strong supporter of education. By his order of 1805, the Paso del Norte district established several public schools for boys twelve to fifteen years of age. Jones (1979) 137-38.
20. Rey to Calvo, February 4, 1836, AHJA f38 s2 p2; Calvo, Chihuahua, to Ponce de León, April 15, 1836, AHJ498 r24; Calvo, Chihuahua, to Ponce de León, February 27, 1836, AHJA f38 s1 p6; Ponce de León, Cobre, to Calvo, April 8, 1836, AHJA f38 s2 p5; List of baptisms, marriages and burials performed in the Mineral de Santa Rita del Cobre May 16–November 1, 1836, by Father Rafel Echeverría, AHJ498 r6.
21. Calvo, Chihuahua, to Rey, March 18, 1836, AHJ498 r24. In the spring of 1836, decisive battles were waged between Texas settlers and the Mexican army, which led to the creation of the Independent Republic of Texas in October of 1836.
22. Pedro Conde, Paso del Norte, to Ponce de León, December 9, 1835; Calvo, Chihuahua, to Ponce de León, December 18, 1835.

Chapter 13

1. Young (1970) 82-83; 189.
2. The San Francisco River was a perfect place for Apache rendezvous. The river, fed by numerous springs and creeks, flowed out of the San Francisco Mountains in southwestern New Mexico into a pleasant valley between the San Francisco and Mogollon ranges to join the Gila River near what is today Clifton, Arizona. The Indians established their rancherías in the valley, but had easy access to mountain retreats in case of trouble. Pearce (1965) 144.
3. Rey, Cobre to Calvo, April 20, 1836, AHJA f38 s2 p6b.
4. Rey, Cobre,to Calvo, May 12, 1836, AHJA f38 s2 p12.

5. Ponce de León, Cobre, to Calvo, May 17, 1836, AHJA f38 s2 p11.

6. Rey, Cobre to Calvo, June 2, 1836, AHJA f38 s2 p14.

7. Rey, Cobre, to Calvo, June 20, 1836, AHJA f38 s2 p21.

8. Daily report of Second Field Division, July 1, 1836, AHJ498 r27.

9. Calvo, Chihuahua to Commander of the second field division in El Cobre, April 8, 1836, AHJ498 r24.

10. Daily report for Second Division, Cobre, June 1, 1836, AHJ498 r24; October 31, 1836, AHJ498 r27. Calvo, Chihuahua to commander of Janos, April 14, 1836, AHJ498 r26; Janos (no signature) to commandante de armas of the western frontier, March 24 1836, AHJ498 r26.

11. Cobre to Calvo, August 4 & 5, 1836, AHJA f38 s2 p26.

12. Whitney (1884) 17; Bartlett (1854) 229.

13. Courcier to Calvo, published in *El Noticioso de Chihuahua*, April 15, 1836, AHJ498 r27.

14. Rafael Tercero to Editors of *El Noticioso de Chihuahua*, April 8, 1836, AHJ498 r27.

15. *El Noticioso de Chihuahua*, October 21.

16. Griffen (1988) 170-171.

17. Daily report for Second Division, Cobre, October 31, 1836, AHJ498 r27. William B. Griffen, in his excellent study of Apaches in the Janos district, concludes that this incident had much to do with the final downfall of Santa Rita del Cobre. Griffen (1988) 72.

Chapter 14

1. Sweeney (1998) 68; Arce, Cobre to Ponce de León, January 26, 1837, AHJ498 r27.

2. Bustamente, Janos to Rey, December 4, 1836, AHJA f38 s1 p16; Arce, Mineral del Cobre to Janos Company, December 23, 1836; AHJA f38 s1 p19a; Arce, Cobre to Calvo, December 24, 1836, AHJA f38 s1 p19b; Calvo, Chihuahua to Ponce de León February 22, 1837, AHJA f40 s1 p8.

3. Arce, Cobre, to Ponce de León, January 3, 1837; AHJ498 r27. In the spring of 1840, Dryden went to Texas, where he reported to President Miranda Lamar on conditions at Santa Fe. Lamar and his supporters wanted to establish the Rio Grande as the boundary of Texas. Dryden returned to Santa Fe in September with instructions to inform the citizens of the advantages of participating with Texas in the fight for independence. In a letter to Lamar, March 10, 1841, Dryden reported that "every American, more than two-thirds of the Mexicans and all the Pueblo Indians are with us heart and soul." In June the ill-fated Texas-Santa Fe expedition set out to extend the Texas boundary. They carried papers appointing William Dryden as one of four commissioners who were to have charge of government affairs in the new territory. When the Texans were defeated and the commission papers discovered, Dryden was accused of being an accomplice in the attempted takeover. He was arrested and imprisoned in the city of Chihuahua. He escaped on the night of January 14,

142

1842, and made his way to California where he died in 1869. Christian (1920) 95-101; Loomis (1958) 264.

4. Arce, Cobre to Ponce de León, January 16, 1837, AHJ498 r27.

5. Daily Report for February 1837, AHJA f40 s1 p11.

6. *El Noticioso de Chihuahua* #104, March 28, 1837, CZ505 r4; Calvo, Chihuahua to Minister of Treasury, March 28, 1837.

7. Almada (1950) 49-57.

8. The Corralitos silver mine flourished, and in May of 1840 the McKnight-Courcier partnership acquired the old Veta Grande mine in the Sierra de Escondida, a few miles southeast of Corralitos. This became the famous Barranca Colorada mine—another money-maker for the partners. Strickland (1965) 267-268. Notice from Luis Zuloaga, Chihuahua, March 7, 1838; Receipt signed by Robert McKnight, Cobre, April 19, 1838, AHJ498 r28.

9. Strickland (1976) 257-286. *El Noticioso de Chihuahua*, May 5, 1837.

10. John C. Cremony (1969) 30- 32.

11. Second Division report, Cobre, June 1, 1837, AHJ498 r27; Public Works, Mineral del Cobre, December 1, 1837, AHJ498 r28.

12. Griffen (1988)175-176; Sweeney (1998) 77-79; Sweeney (1991) 38-40.

13. Griffen (1988) 175-176.

14. Ponce de León, Cobre to José Rodriquez, April 22, 1837, AHJ498 r27; Sweeney (1991) 40.

15. Griffen (1988b) 189.

16. Madrid, Cobre, to Francisco García, May 20, 1838, AHJ498 r28.

17. The last raid on the McKnight/Courcier Santa Rita operation was recounted by United States Boundary Commissioner John Bartlett. He established his headquarters at the abandoned copper mines in 1851 while conducting a survey of the boundary between the United States and Mexico. During the course of his work in the west, Bartlett had the opportunity to hear the story from people who once worked at the Santa Rita mines, and from Lewis Flotte, son-in-law of Robert McKnight. Bartlett (1854) 223; Cutts (1965) 183-184.

18. W. M. Pierson to Secretary of State, Paso del Norte, Mexico, December 6, 1873, in Whitney (1881) 32.

19. Ibid.

20. Strickland (1965) 268; *Diccionario Porrua de Historia, Biografía y Geografía de Mexico,* 541.

21. Wislizenus (1969) 57-58.

22. Spude (1999) 8-38. This is an excellent study of the next period in the history of Santa Rita.

Appendix: The Legend of the Kneeling Nun

Folklore, if so labeled, is an important part of history. Many tales and legends have grown up around the history of Santa Rita del Cobre. The story of the copper mines would not be complete without one of the best known, "The Legend of the Kneeling Nun."

High above the copper mines is a peculiar rock formation, a tall monolith standing in front of an abrupt break in the mountain cliffs. From afar this monolith resembles a veiled figure kneeling in prayer before an altar. This distinctive landmark can be seen for many miles. The Spaniards called the formation "La Bufa" (a comical female singer), a name commonly given to the rocky crests of hilltops in northern Mexico. Lieutenant W.H. Emory, an engineer with General Kearney's expedition, named the formation "Ben Moore Mountain" after a personal friend. Once as tall as the adjacent cliff, the monolith was sometimes called "the needle." In the summer of 1885, however, about thirty feet of the monolith tumbled down, destroying its needle-like appearance.

One day a person with imagination said the formation resembled a woman in nun's habit kneeling before an alter in prayer, and a legend was born. The following version of this legend, published in *The Silver City Enterprise*, June 26, 1885, was written by Walter Sellers, who was a tuberculosis patient at the Fort Bayard hospital for several years and probably spent many hours gazing at the unusual landmark:

The Legend of the Kneeling Nun

This is the tale as they tell it: how in the days of Old
Came the Explorer and Soldier, seeking the glitter of gold;
Robbing and burning and killing, all in the name of the King,
Eye agleam for the honors men to the Conqueror bring.
After them came the Fathers; close on the steps they trod,
Holding aloft the sign of the Faith, chanting the glory of God.
Gentle they were, and tender, healing the wounds of pain
Left by the sword and firebrand by the pitiless hand of Spain.

This is the tale as they tell it: How by the Aztec trail
They builded an Indian Mission, the Knights of the Holy Grail,
Here in the desert they labored, teaching the Truth and the Light
Showing the ways of another race to the savage Sons of Night.
Fairest of all the workers was the Sister Teresa, the Nun,
Touching the Indian children, quickly their hearts she won,
And soon through the desert country, where'er spread the Mission's fame,
Even the gurgling infants were trying to lisp her name.

This is the tale as they tell it: how Diego the Soldier came,
Staggering into the courtyard, weary and sore and lame,
Leagues had he crawled through the desert, seeking a kindly hand;
The last of all his comrades, dead in the new-found land.
Then through the long days of sickness, quietly there by his bed
Watched the Sister Teresa, cooling his fevered head;
And while he raved of his tortures, there through the length of night,
Faithful, kindly, and patient, she watched for the coming of light.

This is the tale as they tell it: how Diego's eyes grew clear,
And gleamed anew with a shining light when the Sister nurse was near.
Hours they walked together—he with his stories of strife
Strange to her quiet seclusion, these tales of the struggle of life.
So did their hearts grow fonder, till ever she bore in her mind
The name of Diego the Soldier, and love to her vows was blind.
Till at last in his arms they found her, her eyes like stars above,
Shining into the depths of her lover's, breathing the Life of Love.

This is the tale as they tell it: how on that fatal day,
Stripped of the garb of her order, they turned the Sister away.
Forth to the desert she wandered and builded an altar of stone;
There she knelt in her sufferings, at last with her God alone.
Then came the storm and darkness, madly the thunder crashed,
Loud rolled the earth in its anger, cruel the lightning flashed;
And oft through the night to the Mission was borne the piteous cry:
"Oh, Madre de Dios, Thy mercy on such as I."

This is the tale as they tell it: how on the coming of light,
There where had been an altar, a mountain had grown in the night;
While before it was kneeling, so saw the Mission flock,
The Sister Teresa of yesterday, turned to eternal rock.
So, in the desert country, through all the length of the days,
Kneeling before her alter, for an erring soul she prays,
And oft, when the storm is raging, they hear her piteous cry:
"Oh, Madre de Dios, Thy mercy on such as I.

Bibliography

Source materials for the story of Santa Rita del Cobre comes primarily from the following archival collections:

University of Texas at El Paso (miocrofilm):

Archivos de la Catedral de Cd. Juárez (ACJ).
Archivos del Ayuntamiento de Cd. Juárez (AAJ).
Archivos del Ayuntamiento de Chihuahua (AAC).
Archivos Historicos de Janos (AHJ).
Carrizal Collection (CZ).
Chihuahua Ayuntamiento Suplementario, Francisco Almada Collection (CAS).
El Fanal (Chihuahua).
El Noticioso (Chihuahua).

State of New Mexico Record Center and Archives:

Mexican Archives of New Mexico (MANM).
Records of the Surveyor General (The Santa Rita del Cobre Grant).
Spanish Archives of New Mexico (SANM).

Archivo General de Indias, Sevilla, Espana:

A.G.I., Audencia de Guadalajara (AGI).

Nettie Lee Benson Latin American Collection, University of Texas, Austin:

Archivos Historicos de Janos (AHJA).

Books

Agricola, Georgius. *De Re Metallica*, trans. by Herbert Clark Hoover and Lou Henry Hoover. New York: Dover Publications Inc., 1950.

Almada, Francisco R. *Gobernadores del Estado de Chihuahua*. Mexico D.F: Imprinta Imprinta de la H. Camara de Deputados, 1950.

_____ *Diccionario de Historia, Geografía y Biografía Chihuahuenses*. Segunda Edicion. Cuidad Juarez: Impresora de Juarez, S. A., 1968.

Archer, Christon. *The Army in Bourbon Mexico 1760–1810*. Albuquerque: University of New Mexico Press, 1977.

_____ *The Birth of Modern Mexico 1780–1824*. El Paso: Scholarly Resources, 2003.

Bakewell, P. J. *Silver Mining and Society in Colonial Mexico: Zacatecas 1546–1700*. Cambridge: The University Press, 1971.

Barrett, Elinore M. *The Mexican Colonial Copper Industry*. Albuquerque: University of New Mexico Press, 1987.

Barry, Louise. *The Beginning of the West*. Topeka: Kansas State Historical Society, 1972.

Bartlett, John Russell. *Personal Narrative of Explorations and Incidents in Texas, New Mexico, California, Sonora and Chihuahua*. New York: Appleton, 1854.

Batman, Richard. *James Pattie's West*. Norman: University of Oklahoma, 1984.

Bolton, Herbert E. *Pageant in the Wilderness: the story of the Escalante Expedition to the Interior Basin, 1776*. Salt Lake City: Utah State Historical Society, 1950. (Reprint from Utah Historical Quarterly, v18 (1950.)

Bowden, J. J. *Spanish and Mexican Land Grants in the Chihuahuan Acquisition*. El Paso: Texas Western Press, 1971.

Brading, David A. *Miners and Merchants in Bourbon Mexico 1763–1810*. Cambridge Latin American Series #10. Cambridge: The University Press, 1971.

Briggs, Walter. *Without Noise of Arms: The 1776 Domínquez- Escalante Search for a Route from Santa Fe to Monterey*. Flagstaff: Northland Press, 1976.

Caballero, Romeo Flores. *Counterrevolution; the Role of the Spaniards in the Independence of Mexico 1804–1838*. Translated by Jaime E. Rodriquez O. Lincoln: University of Nebraska, 1974.

Carroll, H. Bailey and J. Villasana Haggard, trans. *Three New Mexico Chronicles*. The Quivera Society. Albuquerque: University of New Mexico Press, 1942.

Carter, Harvey Lewis. *Dear Old Kit: the Historical Christopher Carson*. Norman: University of Oklahoma, 1968.

Christiansen, Paige W. *The Story of Mining in New Mexico*. Socorro: New Mexico Bureau of Mines and Mineral Resources, 1974.

Clemons, Russell E., Paige W. Christiansen, H. L. James. *Southwestern New Mexico*. Socorro: New Mexico Bureau of Mines and Mineral Resources, 1980.

Cortes, José. *Views From the Apache Frontier: Reports on the Northern Provinces of New Spain, 1799*. Elizabeth A.H. John and John Wheat, editors. Norman: University of Oklahoma Press, 1989.

Cremony, John C. *Life Among the Apaches*. Glorieta: The Rio Grande Press, Inc. 1969. Reprint of 1868 edition.

Cross, Harry Edward. *The Mining Economy of Zacatecas, Mexico in the 19th Century*. Berkley: University of California, PhD dis., 1976.

Cutts, James Madison. *The Conquest of California and New Mexico by the Forces of the United States, in the Years 1846 & 1847*. Albuquerque: Horn and Wallace 1965 reprint.

Diccionario Porrua de Historia, Biografía y Geografía de Mexico. Mexico, D.F.: Editorial Porrua S.A., 1964.

Dobyns, Henry F. *The Apache People*. Tucson: Indian Tribal Series, 1971.

Eggenhofer, Nick. *Wagons, Mules and Men: How the Frontier Moved West*. New York: Hastings House, 1961.

Escudero, José Agustín de. *Noticias Estadísticas del Estado de Chihuahua*. Mexico: J. Ojeda, 1834.

Gibson, George Rutledge. *Journey of a Soldier Under Kearny and Doniphan 1846–1847*. Southwestern Historical Series, v3. Glendale: The Arthur Clark Company, 1935.

Green, Stanley C. *The Mexican Republic: The First Decade 1823–1832*. Pittsburgh: University of Pittsburgh Press, 1987.

Gregg, Josiah. *Commerce of the Praries*. Max L. Moorhead, editor. Norman: University of Oklahoma, 1954.

Gregg, Kate L. editor, *The Road to Santa Fe*. Albuquerque: University of New Mexico Press, 1952.

Griffen, William B., Apaches at War and Peace: The Janos Presidio, 1750–1858, Albuquerque: University of New Mexico Press, 1988.

_____ *Utmost Good Faith: Patterns of Apache-Mexican Hostilities in Northern Chihuahua Border Warfare, 1821–1848*. Albuquerque: University of New Mexico Press, 1988.

Hafen, Leroy. *The Mountain Men and the Fur Trade of the Far West*. 9 vols. Glendale: Arthur H. Clark, 1965–1969.

Halleck, Henry W., ed. *A Collection of Mining Laws of Spain and Mexico*. San Francisco: University of California, 1859.

Hardy, R. W. H. *Travels in the Interior of Mexico in 1825, 1826, 1827, and 1828*. Glorieta: Rio Grande Press, 1977. Reprint of 1829 edition.

Haring, Clarence H. *Spanish Empire in America*. New York: Oxford University Press, 1947.

Humbolt, Alexander De. *Political Essay on the Kingdom of New Spain*. New York: AMS Press, 1966 (reprint of 1811 edition).

Jackson, Donald, ed. *The Journals of Zebulon Montgomery Pike; With Letters and Related Documents*. Norman: University of Oklahoma, 1966.

James, Thomas. *Three Years Among the Mexicans and the Indians*. Glorieta: The Rio Grande
    Press, 1962. Reprint of 1846 edition.

Jenkins, Myra Ellen and Albert N. Schroeder. *A brief History of New Mexico*. Albuquerque:
    University of New Mexico Press, 1974.

Jones, Fayette A. *New Mexico Mines and Minerals*. Santa Fe: Santa Fe, New Mexico Printing
    Company, 1904.

Jones, Oakah L. Jr. *Los Paisanos: Spanish Settlers on the Northern Frontier of New Spain*.
    Norman: University of Oklahoma Press, 1979.

_____ *Nueva Vizcaya: Heartland of the Spanish Frontier*. Albuquerque: University of New Mexico
    Press, 1988.

Joraleman, Ira B. *Copper: the Emcompassing Story of Mankind's First Metal*. Berkeley:
    Howell-North, 1973.

Kessell, John L. *Kiva, Cross and Crown: the Pecos Indians and New Mexico 1540–1840*.
    Washington, D.C: National Park Service, U.S. Department of the Interior, 1979.

Kinnaird, Lawrence. *The Frontiers of New Spain; Nicolas de Lafora's Description 1766–1768*.
    Berkeley: University of California, 1958.

Lockwood, Frank C. *The Apache Indians*. New York: Macmillan, 1938.

Loomis, Noel M. *The Texan-Santa Fe Pioneers*. Norman: University of Oklahoma, 1958.

McCarty, Kieran, ed. *A Frontier Documentary: Sonora and Tucson, 1821–1848*. Tucson:
    University of Arizona Press, 1997.

_____ *Desert Documentary: the Spanish Years 1767–1821*. Tucson: University of Arizona Press,
    1997.

McGaw, William Cochran, *Savage Scene: the Life and Times of James Kirker, Frontier King*. New
    York: Hastings House, 1972.

Meeks, Wilbur T. *The Exchange Media of Colonial Mexico*. New York: Columbia University, 1948.

Meyer, Michael C. and William L. Sherman. *The Course of Mexican History*. New York: Oxford
    University Press, 1979.

Mills, W. W. *Forty Years at El Paso 1858–1898*. El Paso: Carl Hertzog, 1962.

Morgan, Dale. *The West of William H. Ashley 1822–1838*. Denver: Old West Publishing
    Company, 1964.

Moorhead, Max L. *The Apache Frontier*. Norman: University of Oklahoma Press, 1965.

_____ *New Mexico's Royal Road: Trade and Travel on the Chihuahua Trail*. Norman: University of
    Oklahoma Press, 1958.

_____ *The Presidio: Bastion of the Spanish Borderlands*. Norman: University of Oklahoma Press,
    1975.

Navarro García, Luis. *Las Provincias Internas en el Siglo XIX*. Sevilla: Escuela de Estudios
    Hispano-Americanos de Sevilla, 1965.

Pattie, James O. *The Personal Narrative of James O. Pattie*. New York: J. B. Lippincott Company,
    1962. Reprint of 1831 edition.

Quaife, Milo Milton, editor, *Kit Carson's Autobiography*. Lincoln: University of Nebraska, 1966.

Reséndez, Andrés. *Changing National Identities at the Frontier: Texas and New Mexico, 1800–1850*. New York: Cambridge University Press, 2005.

Schroeder, Albert H. *Apache Indians: a Study of the Apache Indians*, part IV. American Indian Ethnohistory. New York: Garland Publishing, Inc., 1974.

Simmons, Marc. *Spanish Government in New Mexico*. Albuquerque: University of New Mexico, 1968.

Smith, Ralph Adam. *Borderlander: The Life of James Kirker, 1793–1852*. Norman: University of Oklahoma Press, 1999.

Sweeney, Edwin R. *Cochise, Chiricahua Apache Chief*. Norman: University of Oklahoma Press, 1991.

_____ *Mangas Coloradas, Chief of the Chiricahua Apaches*. Norman: University of Oklahoma Press, 1998.

Thomas, Alfred Barnaby. *Forgotten Frontiers: a Study of the Spanish Indian Policy of Don Juan Bautista de Anza, Governor of New Mexico 1777–1787*. Norman: University of Oklahoma, 1932.

_____ *Teodoro de Croix and the Northern Frontier of New Spain 1776–1783*. Norman: University of Oklahoma, 1941.

Thrapp, Dan L. *The Conquest of Apachería*. Norman, University of Oklahoma, 1967.

_____ *Victorio and the Mimbres Apaches*. Norman: University of Oklahoma, 1974.

Thurston, Herbert and Donald Attwater, editors, *Butler's Lives of the Saints*. New York: J. P. Kenedy & Sons, 1962.

Voss, Stuart F. *On the Periphery of Nineteenth Century Mexico: Sonora and Sinaloa*. Tucson: University of Arizona Press, 1982.

Walker, Henry Pickering. *The Wagonmasters; High Plains Freighting from the Earliest Days of the Santa Fe Trail to 1880*. Norman: University of Oklahoma Press, 1966.

Wasserman, Mark. *Everyday Life and Politics in 19th Century Mexico*. Albuquerque: University of New Mexico Press, 2000.

Weber, David J., translator and editor. The Extranjeros: Selected Documents from the Mexican Side of the Santa Fe Trail 1825–1828. Santa Fe: Stagecoach Press, 1967.

_____ *The Mexican Frontier, 1821–1846; the American Southwest under Mexico*. Histories of the American Frontier. Albuquerque: University of New Mexico Press, 1982.

_____ *New Spain's Far Northern Frontier: Essays on Spain in the American West, 1540–1821*. Albuquerque: University of New Mexico Press, 1979

_____ *The Taos Trappers: the Fur Trade in the Far Southwest,1540–1846*. Norman: University of Oklahoma, 1968.

West, Robert C. *The Mining Community in Northern New Spain; the Parral Mining District*. Ibero-Americana #30. Berkeley: University of California Press, 1949.

Wislizenus, Adolph. *Memoir of a Tour to Northern Mexico Connected with Colonel Doniphan's Expedition in 1846-1847.* Reprint of 1848 edition. Glorieta: Rio Grande Press, 1969.

Young, Otis E. Jr. *Western Mining.* Norman: University of Oklahoma Press, 1970.

Articles, Pamphlets and Unpublished Material

Allen, R. S. *The Mogollón Mines,* Sixth Annual Edition, 1915.

Almada, Francisco R. "Los Apaches." *Boletín de la Sociedad Chihuahuense de Estudios Historicos,* 2:(June 1939).

_____ "El Archivo de la Comandancia General de las Provincias Internas," *Boletín de la Sociedad Chihuahuense de EstudiosHistoricos,* 1:(July 1938).

Baltosser, W.W. "There Was Action at Santa Rita Sixty Million Years Ago." *Chinorama,* Fall 1968.

Baxter, John O. *Las Carneradas: New Mexico's Sheep Trade to Chihuahua and Durango Before 1846.* Thesis, Master of Arts in History, University of New Mexico, 1982. (This valuable study has now been published by the Historical Society of New Mexico and the University of New Mexico Press).

Bean, Sam. "Early Pioneers." *Rio Grande Republican,* October 26, 1889.

Christian, A. K. "Mirabeau Buonaparte Lamer," *Southwestern Historical Quarterly,* 24:2 (1920).

Faulk, Odie B. "The Presidio: Fortress or Farce," *Journal of the West,* 8:1 (1969).

Foster, Stephen C. "A Sketch of Early Kentucky Pioneers of Los Angeles," *Historical Society of Southern California,* 1:3 (1887).

Golley, Frank B. "James Baird, Early Santa Fe Trader," *Missouri Historical Society Bulletin,* 5:3 (1959).

Griffin, William B. "Apache Indians and the Northern Mexican Peace Establishments," *Southwestern Culture History: Collected Papers In Honor of Albert H. Schroeder.* Published for the Archaeological Society of New Mexico by Ancient City Press, Santa Fe, 1985.

_____ "The Compás: A Chiricahua Apache Family of the Late 18th and Early 19th Centuries," *The American Indian Quarterly,* 7:2 (1983).

Hendricks, Rick. "Spanish Colonial Mining in Southern New Mexico: A Spanish to English Translation of Documents relating to El Paso, the Organ Mountains and Santa Rita del Cobre." *The Mining History Journal,* The Sixth Annual Journal of the Mining History Association, 1999.

Hill, Joseph J. "New Light on Pattie and Southwestern Fur Trade." *Southwestern Historical Quarterly,* 26:4 (April 1923).

Kessell, John L. "Campaigning on the Upper Gila, 1756," *New Mexico Historical Review,* 42:2 (1971).

Kinnaird, Lawrence. "Spanish Tobacco Monopoly in New Mexico 1766-1767," *New Mexico Historical Review,* 21:4 (1946).

Lasky, Samuel G. *Geology and Ore Deposits of the Bayard Area, Central Mining District, New Mexico*, Bulletin #860, Washington: GPO, 1936.

Lindgren, Waldemar, L.C.Graton, C.H.Gordon. *The Ore Deposits of New Mexico*. U.S. Geologocal Survey, Professional Paper #68. Washington: GPO 1959.

Long, William W. *A History of Mining in New Mexico During the Spanish and Mexican Periods*. Thesis, Master of Arts in History, University of New Mexico, 1964.

MacDonald, Donald F., Charles Enzian. *Prospecting and Mining of Copper Ore at Santa Rita, New Mexico*. Washington: GPO U.S. Bureau of Mines Bulletin 107, 1916.

Matson, Daniel S. and Albert H. Schroeder. "Cordero's Description of the Apache 1796," *New Mexico Historical Review*, 32:4 (1957).

Moorhead, Max L. "Spanish Transportation in the Southwest, 1540–1846," *New Mexico Historical Review*, 32:2 (1957).

_____ "The Presidio Supply Problem of New Mexico in the Eighteenth Century." *New Mexico Historical Review*, 36:3 (July 1961).

Park, Joseph F. "Spanish Indian Policy in Northern Mexico 1765–1810." *Arizona and the West*,.4:(1962).

Rickard, T. A. "The Chino Enterprise." *Engineering and Mining Journal*, 116: 18, 19, 23, 26; 117:1. 1923–1924.

Rose, Arthur W., William W. Baltosser. "The Porphyry Copper Deposits at Santa Rita, New Mexico." *Geology of the Porphyry Copper Deposits in Southwestern North America*. Tucson: University of Arizona Press, 1966.

The Santa Rita Native Copper Mines in Grant County, New Mexico. Boston: Algred Mudge & Son, (circa 1882). Huntington Library collection.

Smith, Ralph A. "Apache Plunder Trails Southward, 1831–1840." *New Mexico Historical Review* 37:33 (1962).

Spude, Robert L. "The Santa Rita del Cobre, New Mexico, the Early American Period, 1846–1886," *The Mining History Association Journal*, 1999.

Staples, Anne. "Mexican Mining and Independence: The Saga of Enticing Opportunities." In Archer, Christon I, editor., *Birth of Modern Mexico*. Wilmington: Scholarly Resources, 2003.

Stevens, Robert C. "The Apache Menace in Sonora 1831–1849." *Arizona and the West*, 6:3 (1964).

Strickland, Rex W. "The Birth and Death of a Legend, the Johnson Massacre of 1837," *Arizona and the West*, 18:3 (1976).

_____ "Lewis Dutton," in Hafen, LeRoy R, editor, *The Mountain Men and the Fur Trade of the Far West*, 9:147-152. Glendale: The Arthur H. Clark Company, 1965.

_____ "Robert McKnight," in Hafen, *The Mountain Men and the Fur Trade of the Far West*, 9:259-268. Glendale: The Arthur H. Clark Company, 1965.

Sully, John M. "The Chino Copper Company: the Story of the Santa Rita del Cobre Grant and its Development," *Resources and Industries of the Sunshine State*. Nd.

_____ "The Santa Rita del Cobre Grant." *The Silver City Enterprise*, December 1, 1933.

Timmons, W. H. "The El Paso Area in the Mexican Period, 1821–1848, *Southwestern Historical Quarterly*, 74:1 (1980).

Walker, Billy D. "Copper Genesis: The Early Years of Santa Rita del Cobre." *New Mexico Historical Review*, 54:1, January 1979.

Warner, J. J. "Reminisences of Early California from 1831 to 1846," *Historical Society of Southern California*, Annual Publication 7, 1907–1908.

Willard, Rowland. "Inland Trade with New Mexico," Appendices to *The Personal Narrative of James Ohio Pattie*. Cincinnati: John H. Wood, 1831.

Index

Jáurequi, Francisco, 37-38, 55-56, 61

Justiniani, Cayetano, 89-94, 96, 100-104, 139n4

Kennecott Copper Corporation, 9, 125

Kirker, James, 57, 71, 76, 78, 121-122, 137n14, 139n4

Kirker, José, 57

Kneeling Nun, 22, 123-124, 128n5, 144-145

La Bufa. See Kneeling Nun

labor, 24, 26, 31, 37, 49, 56, 69, 88, 102, 104; Apaches work at mines, 24, 83, 86. See also convicts, penal colony

Madero, Jose Isidro, 78

Madrid, Todocio, 120

Madrid, José, 85

Magruder, John, 133n4

Mangas Springs, 58, 84

Mano Mocha, 59-60, 64-65, 69-70, 77, 82-84, 107

marriages, 45, 141n20

Matamoros, 95, 98, 111, 140n2

McKnight, Robert, 73-78, 81, 84-86, 88-89, 91, 93-94, 98-99, 101-105, 108-111, 115, 117-118, 121-122, 136n1, 143n8

McKnight, John, 98, 137n8

Meras, Francisco, 37

Mexican War for Independence, 57, 60-63, 66, 80, 95

Mexico City, 64, 78

Miera y Pacheco, Bernardo, 16-17, 127n4

military operations (Spanish, Mexican), 16, 19, 21, 27, 30-31, 42, 45, 47, 58-59, 62-63, 65, 78, 80-82, 85-86, 89-90, 92, 94, 96, 99, 103, 105, 110, 113, 116, 119-121

militia, civil, 16, 18, 30, 58, 81-82, 84, 89, 100-101, 105, 115

Mimbres Mountains, 15-17, 58, 82, 131n6

Mimbres River (valley), 24-25, 41, 59-60, 64, 70-71, 75, 77, 83-84, 86, 95, 107, 116-117

mint, 11, 35, 52, 62-65, 78, 81, 88, 101-102, 111, 113, 116-117

Mogollon Mountains, 58-60, 64, 82, 85, 127n5, 141n2

Molinares, Simon, 57

Moreno, Pablo, 115

mule trains, 12, 24, 26, 30-31, 33, 38, 40, 42-43, 48-50, 52, 58, 61-64, 68, 70, 73, 75, 85, 94, 96, 121, 129n2

Narbona, Antonio, 68

Nava, Pedro de, Commandant General, 24-25, 27

Navajo, 65, 109

Negrita Mountains., 58

New Mexico, 9-11, 13, 15-17, 19, 25, 27, 30-31, 39, 42-43, 48, 50-51, 53, 60-61, 67-68, 70, 73, 76, 81-82, 86, 93, 95, 112, 122, 124-125, 127n4, 128n11, 129n26, 136n8, 139n4

New Spain, 11, 15-16, 21, 48-49, 51, 62, 67, 128n11, 129n22

Nogales Canyon, 57

Obeso, Diego, 33, 43

Ochoa, José Manuel. 26

Ojo de San Vicente, 25, 28, 59, 129n26

Ojo de Vaca, 33

Opata Indians, 26, 129n19

Orcacitas, Francisco, 102

Pachitejú, 33, 59-60, 81, 95, 131n6

Parral, 37, 139n4

Paso de Todos los Santos, 17, 127n5

Pattie, Sylvester, 66-69, 71-72, 75, 77

Pattie, James O., 67-72

peace negotiations, 18-19, 57-58, 60, 65, 69-70, 75. 82-84, 88-89, 92-93, 107-109, 113-114, 121-122

peace establishments, 19, 21, 24, 33, 41, 57-59, 65, 77, 80

peaceful interludes, 58,60, 64, 69-70, 77, 83

Pecos, 127n4

Peña, José, 113

penal colony, 89-90, 99-100, 102, 117, 119

Phelps Dodge Corporation, 10

Pike, Zebulon Montgomery, 48

Pisago Cabazon, 82, 93, 107-108, 113-115, 119

Platón, Fr. Augustín Gómez, 45, 52, 133n20

Pluma, Antonio, 83

Ponce de León, Mariano, 104-105, 110, 113-115, 117, 120

Pratte, Bernardo, 68

presidios, 13, 18-19, 21, 24-26, 30-31, 33, 39, 41-42, 45, 52, 57-59, 61, 65, 76-78, 80-86, 88-92, 94-96, 99-106, 108-110, 115-117, 119, 127, 131n4

priest, 29, 44-45, 99, 105

Pryor, Nathaniel, 71

Queretaro, 64

Ramirez, Alexander, 138n16, 141n13

Ramos de Verea, Pedro, 24-26

Rancho la Pastoría, 28

Real del Cobre (mining camp), 11, 26, 28-29, 35-36, 38-47, 50, 55, 57-60, 64, 75, 81, 84, 86, 88-89, 98-99, 103, 105, 107, 114-115; defense of, 40-42, 50, 81, 95, 117. See Apache raids on Santa Rita del Cobre

Rey, Mariano Rodríquez, 96, 104-105, 108-109

Rio Grande, 15-17, 31, 48, 68, 142n3

Rodriguez, Juan, 83

Ronquillo, José Ignacio, 82

Royal Treasury. See mint

Rueda, Diego Gonzáles, 56-57

Ruiz, Nicolas, 83

Salcedo, Nemecio, Commandant General, 27-28, 32, 35, 42-43, 45, 47-48, 51-52, 57-59, 129n22, 141n19

San Buenaventura, 21, 25, 31, 33, 41, 58, 82, 91, 94, 104-105, 109-110

San Elizario, 59, 137n11, 137n12

San Francisco River (valley), 82, 85, 107-108, 141n2

San José mine, 56, 133n4

San Luis Potosi, 64

San Mateo Mountains., 65

San Miguel del Vado, N.M., 94

San Miguel River, 33

Santa Anna, 95

Santa Fe, 25, 31, 43, 61, 68-69, 72, 74, 76-77, 98, 101

Santa Lucia Springs, 58, 60, 82

Santa Rita de Casia, 36

Santa Rita del Cobre mines, 9-12, 17, 20-21, 23, 25-26; claims, 27, 36, 55-56, 76, 101; development, 28, 31, 33-34, 47-48, 53; quality of copper, 35; markets, 12, 23, 35, 51-52, 60-65, 78-79, 111; production, 26, 35-38, 46-48, 51-53, 60-64, 66, 70, 74, 78-79 ,81, 111, 122, 125; smelting, 47-51; supplies, 24, 27, 31, 40, 42, 64, 75, 81, 84, 91, 140n2. See also Real del Cobre

Santa Rita del Cobre mining camp. See Real del Cobre

Santa Rita Gold Mine, 37-38

Santa Rita Presidio, 90-91, 94-96, 99-100, 102, 104-110, 115-117, 119, 131n3

Santisima Trinidad del Oro, 25, 36, 40; see also gold at Santa Rita del Cobre